T0074799

THE GLOBALIZATION OF WHEAT

INTERSECTIONS: HISTORIES OF ENVIRONMENT, SCIENCE, AND TECHNOLOGY IN THE ANTHROPOCENE

SARAH ELKIND AND FINN ARNE JØRGENSEN, EDITORS

MARCI R. BARANSKI
THE GLOBALIZATION
OF WHEAT

A CRITICAL HISTORY
OF THE GREEN REVOLUTION

UNIVERSITY OF PITTSBURGH PRESS

Published by the University of Pittsburgh Press, Pittsburgh, Pa., 15260
Copyright © 2022, University of Pittsburgh Press
All rights reserved
Manufactured in the United States of America
Printed on acid-free paper
10 9 8 7 6 5 4 3 2 1

Library of Congress Cataloging-in-Publication Data

Names: Baranski, Marci, author.
Title: The globalization of wheat : a critical history of the green
    revolution / Marci R. Baranski.
Description: Pittsburgh, PA : University of Pittsburgh Press, [2022] |
    Series: Intersections. Histories of environment, science, and technology
    in the anthropocene | Includes bibliographical references and index.
Identifiers: LCCN 2022030115 | ISBN 9780822947349 (cloth ; acid-free paper)
    | ISBN 9780822989066 (ebook ; acid-free paper)
Subjects: LCSH: Borlaug, Norman E. (Norman Ernest), 1914-2009. | Borlaug,
    Norman E. (Norman Ernest), 1914-2009 Wheat in the Third World. |
    Wheat--Developing countries--History. | Green revolution--History.
Classification: LCC SB191.W5 B348 2022 | DDC
    633.1/1091724--dc23/eng/20220722
LC record available at https://lccn.loc.gov/2022030115

ISBN 13: 978-0-8229-4734-9
ISBN 10: 0-8229-4734-X

Cover design by Melissa Dias-Mandoly

To my parents, Mark and Nanette

CONTENTS

ACKNOWLEDGMENTS

This book exists only because of the intellectual and personal generosity of the community of scholars and scientists in my life. I am especially thankful to my friends and colleagues at Arizona State University—the scrappy community of historians, philosophers, ethicists, economists, and ecologists—who read my drafts and listened to my talks. I am permanently indebted to Jane Maienschein and Clark Miller for creating this unique community of scholars from whom I've learned so much. I have endless thanks for Jessica Ranney and the administrative staff at ASU for helping me navigate paperwork and getting me to the right places. Ann Kinzig is not only brilliant but also a wonderful mentor who guided, challenged, and supported me through the highs and lows of my research. Hallie Eakin, Dan Sarewitz, and Jamey Wetmore have all inspired me to reach higher and work harder. Prem Narain Mathur welcomed me into Bioversity International's India office, helped me navigate New Delhi, and introduced me to the right people in Indian agricultural science. I am also beholden to the other members of the Bioversity International New Delhi office during my time there, including Bhuwon Sthapit, Sarika Mittra, Madan Kanthaganj, and the administrative staff in Delhi and Rome. I'm also especially grateful to Mark Largent for guiding me toward ASU, science policy, and the history of science.

The National Security Education Program Boren Fellowship, Rocke-feller Archive Center, and the ASU Center for Biology all provided finan-cial support that made this research possible. My archival research was guided by the helpful librarians and archivists at the Indian Agricultural Research Institute (IARI) Library, Iowa State University Library, and the Rockefeller Archive Center, especially Lee Hiltzik. At IARI Pusa, I thank Drs. I. S. Solanki and D. U. M. Rao for their hospitality and assistance. At the Directorate of Wheat Research, Dr. Satyavir Singh and Dr. Dinesh Kumar were the most thoughtful hosts. Thank you Dr. Achla Sharma and Dr. N. S. Bains at Punjab Agricultural University for your kindness and insights. At IARI New Delhi, Dr. Malvika Dadlani, Dr. Anju Ma-hindru, and Dr. Rajbir Yadav all helped me navigate the historic IARI campus. Dr. Yadav in particular answered many, many questions I had about wide adaptation and the Indian wheat research system, and also tolerated an unnecessarily early taxi with me from Pusa to Patna. Thank you also to the many researchers who spent time with me and helped me understand the complexities of the Indian wheat research system.

Many people have played a special role in this book's development. My anonymous reviewers were invaluable in helping me clarify my nar-rative and identify additional areas of research. Audra Wolfe provided insightful feedback that guided my final revisions, and Regina Higgins provided editorial and formatting support. Abby Collier at the Univer-sity of Pittsburgh Press got this manuscript finally over the finish line. Folks from the agricultural history and agricultural research commu-nity also helped me refine my writing and analysis, including Barbara Kimmelman, Jonathan Harwood, Mauricio Bellon, Derek Byerlee, Sal-vatore Ceccarelli, and members of the Ernst-Struengmann Forum on Agrobiodiversity in the 21st Century, especially Jacob van Etten, Glenn Davis Stone, and Karl Zimmerer. Mary Ollenburger in particular helped me bring my analysis further into the twenty-first century.

This has been a long journey and I am eternally thankful for every friend and colleague who provided feedback and encouragement, but es-pecially Kate MacCord, Paige Madison, Erick Peirson, and Steve Elliott. Thank you to my family for your support as I bounce across time zones. Jess, thank you for hosting me in Ames; I can't tell you how lucky that was! Michael, Tim, and Kevin—you each played an important part of my life during different phases of this book. Michael, we shared those exciting first years of graduate school together as I developed the concept that became this book. Tim, you emotionally supported me through my fieldwork and several setbacks on the way to publishing this manuscript. Kevin, you mostly get to reap the benefits(?) of dating an author. Thank you for prompting me to find a way back to Asia and supporting me through the final edits. Finally, thank you to the cats—Fluffins, Mixie, Boonchok, Isaac, and Nikolai—who kept me company during many long days and nights of research and editing.

THE GLOBALIZATION OF WHEAT

INTRODUCTION

Although wheat in general appears to be a plant with varieties which are comparatively specialized, nevertheless, in many ecological types there is observed a high degree of ecological plasticity.

—N. I. VAVILOV

As a graduate student in the early 2010s, I wanted to understand how agricultural scientists in India were planning to adapt wheat and rice farming to a hotter, drier climate. During my first research trip, I visited several remote research stations in northern India. I experienced many versions of the same standard tour of the field trials and laboratories. My tour guides boasted of the achievements of their station's scientists, such as a popular crop variety or piece of machinery. While examining collections of crop varieties, I heard scientists again and again refer to prized varieties possessing the ideal property of "wide adaptation." When I asked one scientist what this meant, he said it meant a variety that excels under any condition. Of course, my colleagues in ecology (including my own dissertation advisor) would immediately call this preposterous—few species excel under all or even most conditions. So instead of studying climate change adaptation in India, I decided to look at the concept of wide adaptation. I wondered: why were agricultural scientists still talking about wide adaptation, something I thought had been discredited after the Green Revolution?[1] I soon realized the reason I was hearing about wide adaptation at these research stations in India. All paths led back to Norman Borlaug, the Nobel Peace Prize laureate and American agricultural scientist.

I was familiar with Borlaug because for several years I idolized him. At thirteen years old, I decided to become a biochemist after reading about Golden Rice—rice that has been genetically engineered to produce beta-carotene. I studied biochemistry at Michigan State University and learned about Borlaug's story of feeding the world through science. I even had a magazine photo of him pasted onto the back of my college diary. In the parlance of Charles C. Mann, I followed the "wizard" path of using science and technology to improve society. My technological optimism about agriculture quickly faded, however, as I started taking social science and history classes. I took an internship in Bangladesh to study the adoption of hybrid rice by women and coastal farmers. There I saw how agricultural technologies could either empower or constrain farmers under different contexts, and how climate change could exacerbate the social inequalities that always mediate technology adoption. This led me to my graduate studies in biology and society at Arizona State University, where I reencountered Borlaug's work from a more interdisciplinary and critical perspective. As you will see, much of this book is about Borlaug and the complicated memories we hold of him and the Green Revolution.

In the 1950s and 1960s, Borlaug combined a few unique genetic traits into new varieties of wheat and started a program that would revolutionize agriculture. He used novel breeding methods to combine the traits of photoperiod insensitivity, dwarfing, and rust resistance in wheat while working for the Rockefeller Foundation in Mexico. He bred wheat varieties with a wide growing range that also responded to high levels of fertilizers, which Borlaug called "surprisingly broad adaptation."[2] Scientists define *adaptation* as the fit of an organism to an environment or geography. Borlaug realized that these varieties had high yields in locations ranging from Mexico to Argentina and Kenya to India. Many scientists viewed Borlaug's breeding work with skepticism, but Borlaug convinced the Rockefeller Foundation (RF) to use broad adaptation (also called wide adaptation) as a purposeful breeding strategy for wheat and other crops. In this model, one international research center could provide varieties to countries with limited agricultural research capacities. The RF's efforts in crops other than wheat had varying levels of success. But the rapid spread of particular wheat varieties through South America, the Middle East, and South Asia and the subsequent increase in global wheat (and later rice) production was proclaimed the "Green Revolution."

The Green Revolution did not happen in a political vacuum: it was the result of scientific advances converging with political motives. During the Cold War, the United States believed that nonaligned countries such as India would be more receptive to food aid and scientific assistance rather than direct diplomatic intervention.[3] Thus, the US government facilitated the RF to support agricultural research in several countries. This put Borlaug in a favorable position to spread both the new wheat varieties and his radical concept of wide adaptation.

Borlaug's research program on wide adaptation emerged in tandem with his focus on breeding crop varieties for ideal agronomic conditions (high fertility and controlled irrigation). Borlaug fervently promoted research and testing of wheat *only* under high fertility conditions. He believed that fertilizers would soon become widely available and affordable around the world, as they did in the United States during his youth on a farm in Iowa. After entering his wheat varieties in several international yield tests, he soon began claiming that these varieties could out-yield almost any local variety, fertilized or unfertilized.

Unfortunately, Borlaug's exclusive focus on wide adaptation and high levels of fertilizers led him down a narrow ideological path of believing his methods were the *only* solution to hunger. Borlaug urged foreign scientists and governments to quickly adopt his wheat varieties and synthetic fertilizers, especially in the Middle East, India, and Pakistan. Borlaug and the RF also helped install a centralized research system in India that would initially target farmers with access to irrigation and fertilizers. Borlaug claimed that the focus on wealthier farmers was justified by the urgency of the food problem. He and other scientists assured politicians that the new technologies would spread to benefit all farmers (this is called "technological spillover"). He argued that while the RF would initially focus on irrigated, highly fertilized land, a widely adapted crop was not limited by these conditions.

In my research I found that scientists used the doctrine of wide adaptation to justify overlooking the agroclimatic and socioeconomic diversity inherent to farming. Research was aimed at the ideal, "innovative" farmer who managed a commercial farm, and scientists assumed the benefits would trickle down to marginal and smallholder farmers. Scientists called technologies, especially seeds, "scale neutral" and assumed they would benefit both small and larger farmers (that is, no economy of scale was needed). The Green Revolution, however, has failed to produce the promised spillover effects and has mostly benefited larger farmers who commercialized their farms. In other words, while the seed is an enticing technological solution, the socioeconomic and agroclimatic contexts of farming create an uneven playing field.

Unfortunately, many agricultural research and development practitioners still believe that seed and fertilizer technologies are scale neutral.[4] This mostly due to the deeply internalized narrative of the Green Revolution, according to which scientists and administrators used scale neutrality to counter criticisms of technological and economic inequality. Practitioners also use the scale-neutral argument to eschew the messier work of agricultural development that involves social and political change. Despite recognition within the social sciences that "it is not possible to disassociate the social and economic effects of the technology from the social system in which it is functioning," the belief in scale neutrality has persisted in international development.[5] Meanwhile, the Green Revolution still has daily consequences for the millions of smallholder farmers in the world who have been left behind.

In reaction to an episode of the *American Experience* documentary series on Borlaug, journalist Justin Cremer wrote that "historian Tore Olsson suggests that widespread famine and hunger still remain not because of a lack of food but because of inequality, class and poverty."[6] Cremer argued, "This may indeed all be true, but the way the filmmakers seem to lay these enormous problems at Borlaug's feet is unfair."[7] I agree—we cannot blame Borlaug for the shortcomings of the modern food system (and I believe that was not Olsson's intention). But we can ask ourselves, who is served by our upholding Borlaug as the paragon of agricultural science? Who benefits from the idea that the solution to hunger is more technology, and that agricultural technologies are scale neutral?

Many decades of social science research have shown that agricultural technologies are not scale neutral and that hunger is more related to food access than to agricultural production. Yet the major agricultural developmental organizations today still follow the Green Revolution recipe of top-down technological development that has largely ignored smallholder farmers. Smallholder farmers produce over half of the world's calories, yet a recent analysis of agriculture research publications found that over 95 percent of the studies "were not relevant to the needs of smallholders and their families."[8] Today's agricultural development models aim to "scale up" agricultural technologies. Yet there is a fundamental incompatibility between the "scalability" of technologies and "the enormous heterogeneity of contexts faced by smallholder farmers."[9] Though the term "wide adaptation" is less prominent today than during the Green Revolution, mainstream agricultural development organizations still rely heavily on the wide adaptation concept to support the claim that technologies can be unproblematically scaled across diverse environmental and socioecological conditions.

THE CONTESTED HISTORY OF THE GREEN REVOLUTION

Given its centrality to this book, let us examine the Green Revolution for a moment. The Green Revolution was the convergence of political and

scientific resources aimed at increasing the yields of cereal crops starting in the 1940s and through the 1970s. It is commonly summarized by
the following story: the Rockefeller Foundation supported scientists to
develop improved and high-yielding crop varieties that were adopted in
Latin and South America and South and Southeast Asia.[10] These varieties
of maize, rice, and wheat increased yields and food production, saving
millions of people from a Malthusian catastrophe of overpopulation and
underproduction. A vague claim about reducing poverty typically ac-
companies this narrative.

Most of this narrative is highly contested. To start with the uncon-
tested, certain countries did increase production of wheat and rice in the
1960s and 1970s. But to pin this increase entirely to new crop varieties is
misguided. Improved crop varieties alone seldom substantially increase
yields. The introduction of new wheat varieties in India, for example, was
accompanied by state support for infrastructure like irrigation, in subsidies for fertilizer and other inputs, in new government markets, and in
the mobilization of a large network of agricultural extension workers to
train farmers on new agronomic techniques. Kapil Subramanian argued
that investments in irrigation led to a higher annual growth rate in food
crop yields *before* the introduction of new seeds in India.[11] Richa Kumar
posited that the production increase in wheat was largely due to market incentives, and had a different crop been targeted, the same outcome
would have occurred, thus dispelling the "miracle" technology narrative
of the Green Revolution.[12]

Even with infrastructural support, the Green Revolution's successes
were concentrated among wealthier farmers. Literature starting in the
late 1960s has shown the unequal impacts of the Green Revolution.[13]
These were in part due to deliberate targeting of technologies to favorable
areas.[14] Wealthier farmers were able to adopt riskier new technologies
because they usually had access to irrigation, which mitigates the risk
of crop failure. The biology of wheat is also material: Green Revolution
wheat varieties required specific agronomic conditions and high rates of
fertilizers typically achieved only by researchers and wealthier farmers.[15]
Considerable ink has been spilled over whether higher wheat yields on
large farms could boost rural economies through increased wages and
lower food prices, with no clear consensus. Yet a review of literature by
the agricultural economist Donald Freebairn found that in studies of the
economic impacts of Green Revolution that had a conclusion, 80 percent
found that income inequality increased in Green Revolution–affected
areas.[16] Many critics of the Green Revolution have highlighted this apparent contradiction: while wheat production increased in some regions,
inequalities increased and poverty and malnutrition persisted.

Next, claims that the Green Revolution prevented widespread famine in the second half of the twentieth century are exaggerated.[17] Hunger levels did fall globally between 1965 and 2000 for a variety of reasons, including the Green Revolution.[18] Increased production of cereal crops contributed to lower food prices, which is important to food security. But the common Green Revolution narrative describes India on the brink of famine in 1966 due to two consecutive droughts during the rainy season. Historian Nick Cullather showed that American diplomats constructed this narrative to push their own interests. Although there was a famine in Bihar during 1966, it was caused not by food shortages but rather a crash in rural incomes from a poor crop of jute and sugarcane.[19] The Indian government acted decisively to distribute food to hungry Biharis, resulting in relatively few deaths: less than 2,500 compared to over 2 million who died during the Bengal famine in the 1940s.[20] The idea of India perpetually teetering on the brink of starvation was a Western fantasy more than reality. As scholars have highlighted the role of British colonialism in exacerbating deaths and distress during famines in India, we now understand that famine often has more to do with socioeconomic entitlements and political choices than with the availability of food.[21] Nonetheless, chronic malnutrition remains pervasive globally and especially in South Asia. It is clear from the prevalence of both hunger and poverty in Green Revolution–affected countries that "the Green Revolution, as a story about technological triumph over hunger . . . ignores the question of whether increased yields led to reduced hunger."[22] India is self-sufficient in grain production, yet one-third of the world's malnourished children reside there and the rate of severe wasting in children is so high in South Asia it is considered a public health emergency.[23]

Finally, while poverty rates have declined worldwide, agricultural growth is only one of several contributors to poverty reduction. In India, the prevalence of poverty seems correlated with adoption of Green Revolution varieties at some point in time, but there is little correlation between adoption of agricultural technologies and poverty *reduction*. Raju Das showed that some states in India have reduced poverty despite low adoption of Green Revolution technologies and some high-adoption states have made less progress toward reducing poverty.[24] In early 2021, farmers in India protested new farm laws that they feared would compound their existing stressors of agricultural debt and bankruptcy.[25] Literature from the social sciences shows that in current contexts, higher crop yields rarely improve poverty levels for smallholder farmers.[26] It is also worth noting that the Green Revolution "bypassed" most of Africa, largely because African countries did not have the same political support and infrastructure for agricultural development and because farmers in sub-Saharan Africa tend to be smallholders in very diverse agricultural environments.

SMALLHOLDER FARMERS

Much of my analysis of the Green Revolution and its consequences focuses on smallholder farmers. This is for two reasons: first, I believe the agricultural development community has an ethical obligation to correct the deficiencies of the Green Revolution and ensure that future efforts do not further contribute to inequity, and second, we must engage smallholder farmers to sustainably transform agricultural systems. Agriculture is the main driver of deforestation, land degradation, and biodiversity loss. This is partly due to the low productivity of smallholder farming systems leading to the expansion of agricultural land (although small farms can be highly productive and large farms also contribute to ecological problems). While these farms are small, there are many of them. A recent study found that "smallholder-dominated systems are home to more than 380 million farming households, make up roughly 30% of the agricultural land and produce more than 70% of the food calories produced in these regions, and are responsible for more than half of the food calories produced globally."[27] In India, almost half the farming area is made up of marginal and small farmers with average farm size of 0.6 hectares (1.5 acres).[28] Despite the prevalence of smallholder farming throughout the developing world, the agricultural research community has for the most part ignored farming under these diverse circumstances.[29]

Why have smallholder farmers not benefited from agricultural development? Although there are historical examples of technology development aimed at peasant farmers, these models are overlooked by modern development practitioners. Both Japan and Bavaria, Germany, used decentralized crop breeding programs to adapt varieties to the conditions of local smallholders in the early 1900s.[30] Borlaug's wheat actually descended from the Japanese wheat variety Norin 10, developed by Gonjiro Inazuka. But Green Revolution scientists imported the technology without the context of its development, and instead chose a centralized breeding model. In modern times, there are also examples of developing technologies for poorer farmers, such as the case of research on rainfed rice in Bangladesh.[31] Technologies can help smallholder farmers, but they often must be accompanied by socioeconomic factors such as land reform, strong farmer groups and cooperatives, access to markets, and improvements in education and extension. These efforts are more difficult than distributing seed and fertilizer.

Promoters of the Green Revolution often accuse critics of romanticizing smallholder agriculture and peasantry, arguing that in the absence of the Green Revolution, millions would still be hungry and poor. I argue, however, that we should aim primarily to reduce poverty and food insecurity and that technological improvement should be considered one of many tools, rather than the primary driver of that outcome. The question is not whether increase in food production is good or not, but whether it led to the desired outcomes—and this requires closer attention to the improvement of social and economic conditions.[32]

The way we envision the past matters to how we envision the future. In much of today's agricultural development, interventions are centered on top-down technological solutions. Several international organizations aim to reproduce last century's Green Revolution through high-input, high-production agriculture, while also tacking on the goals of poverty reduction and food security. Yet these organizations have not addressed how their technology-driven agenda would result in a more equitable outcome rather than replicating the inequities of the Green Revolution. These organizations seek to extend the Green Revolution while claiming that they will avoid its negative impacts, such as environmental degradation and inequality. However, none of these efforts substantially show *how* they will avoid this or *how* higher yields will reduce hunger and poverty.[33]

The Green Revolution was a massive, global transformation in how food was produced, resulting in increased production and lower food prices without expansion of agricultural land. In this book, however, I confront the fact that the Green Revolution largely bypassed small and marginal farms. Addressing historical and current inequities requires a deep dive into one of the most important assumptions of the Green Revolution: that crops can be widely adapted across both physical and agroecological environments. I argue that this flawed premise has led to the concentration of research efforts on favorable agroecosystems to the detriment of smallholder farmers working in highly variable, often marginal environments for agriculture.

NEW CRITIQUES OF THE GREEN REVOLUTION

Histories of the Green Revolution tend to take one of two forms: the "wizardly" narrative, highlighting the achievements of scientists against hunger, and the "prophet"-like critiques, which come mostly from the humanities and social sciences. I find the wizardly set of histories largely unsatisfactory.[34] These narratives are usually written by nonhistorian authors who mostly ignore the existing scholarship around the history of the Green Revolution. These accounts tend to overlook the sociopolitical context of the Green Revolution and focus on scientists who resolved to fight hunger and the challenges they overcame, and celebrating the outcome. They also gloss over the process of making history—which is that histories are often contested.

In this book I take a more critical view of agricultural scientists and of the Green Revolution. *Critical* does not mean negative; it means examining assumptions. We encourage "critical thinking" in our society so that people are not fooled by charlatans. When critiques focus on our preferred scientific narratives, we should also try to put our defenses aside and evaluate the evidence presented. The COVID-19 pandemic showed how the prevailing paradigm of surface transmission of disease was due for an update centered on airborne transmission. Science is not a static subject, and neither is our interpretation of history.

This critical approach is by no means unprecedented. Recent scholarship has argued that the Green Revolution should be viewed as a long period of agricultural improvement rather than as a radical leap in productivity due to new crop varieties.[35] These recent works also highlight the role of non-American scientists, bureaucrats, and citizens in agricultural development. Building on these insights, I will show that the Green Revolution was not a linear transfer of technology but rather a result of coinciding sociopolitical forces including the Cold War, technological optimism, the rising status of plant breeders, new institutions in international aid and development, and changing agricultural policies. I will specifically examine the issue of wide adaptation in Green Revolution wheat. Taking a deep dive into Borlaug's wheat program, I will show that widely adapted wheat was more contested and less evidence-based than we might assume.

Recent histories of the Indian Green Revolution corroborate and complement each other, contradicting Borlaug's key claim that his varieties could outperform tall Indian varieties. In addition to Subramanian's argument that new seed technologies contributed little to food security in India, Glenn Davis Stone showed that overall cereal production in India increased at a linear rate from 1950 to 2000, showing no evidence of the exponential growth claimed by supporters of the Green Revolution.[36] In other words, higher wheat production was offset by lower production of so-called coarse cereal crops (such as millet, sorghum, and oats). Stone summarized, "The legendary wheat-field triumphs came from financial incentives, irrigation, and the return of the rains, and they came at the expense of more important food crops."[37] While these recent studies have shaken the foundations of the Green Revolution narrative, the Green Revolution myths remain widespread in the agricultural development community.

Historian Jonathan Harwood has the most prolific recent bibliography on the Green Revolution, bridging the history of science with development studies to explore not just what lessons can be learned from the Green Revolution but *how* they can be used for science policy. His 2012 book, *Europe's Green Revolution and Its Successors*, which focuses on plant breeding in Germany, provides a brief but illuminating overview of the Asian Green Revolution and its shortcomings, how Green Revolution scientists responded to critiques, and reviews several "peasant-friendly" forms of agricultural research. Harwood also highlighted that "the original GR [Green Revolution] was not primarily driven by a concern to alleviate hunger but rather by the aim of promoting the use of commercial inputs among small farmers," and brought much-needed exposure to the role of fertilizer companies in the Green Revolution.[38] He has also discussed how Green Revolution scientists responded to criticisms, which is a much-needed contribution to the literature.[39]

Unfortunately, critiques of the Green Revolution tend to diverge by discipline. Natural scientists and economists tend to uphold the claim that "long-term improvement in agricultural productivity has helped ward off the Malthusian catastrophe predicted in the 1960s" while recognizing some of the ecological and social impacts of the Green Revolution.[40] Social scientists and humanities scholars, along with agroecological scientists, view the Green Revolution as disastrous to ecologies and economies of rural areas and warn that without changes, these trends will continue in the "new Green Revolution" agenda. Despite the fact that the both the history and future trajectory of the Green Revolution are highly contested, people continue to extract "lessons" from history to achieve certain political goals.[41]

NARRATIVES AND TRAJECTORIES IN AGRICULTURAL RESEARCH

Scholarship over the past decade has discussed the importance of the Green Revolution narrative in agricultural research and development.[42] Narratives can simplify a complex issue, but they can also condense a historical event into a preferred policy. Green Revolution narratives are still incredibly pervasive and influential in agricultural research, and given the myths about the Green Revolution that recent scholarship has dispelled, reliance on these narratives presents several challenges to equitable research and development.

Narratives of the Green Revolution, typically told by agricultural scientists or administrators with little regard for historical scholarship, uphold the story of "miracle" plant varieties as central to agricultural development. The Green Revolution narrative emphasizes that improved crop productivity led to reduced hunger and poverty. Critical to this narrative is the "political myth" that the Green Revolution averted famine in the 1960s.[43] While narratives around the Green Revolution also refer to irrigation, fertilizer, and mechanization as supporting players, the role of political, economic, and social change are typically left out. Philanthropic institutions are overemphasized, and corporations and governments are deemphasized. Roger Pielke Jr. and Bjorn-Ola Linnér referred to this narrative when they wrote, "Famine averted by the intervention of scientific genius is a much more straightforward narrative than a famine-free story of incremental, accumulating, multi-factor progress in local agricultural production due to a complex tapestry of societal and political actors."[44] The simple narrative of improved seeds prevailed over the complex political, social, and institutional changes that facilitated the Green Revolution.

The Green Revolution narrative shapes how we conceptualize agricultural research. This narrative holds that "modern agricultural technology to maximise crop yield is . . . the essential weapon in the battle against food insecurity, hunger and starvation."[45] Therefore, philanthropies and foreign aid focus on plant breeding as a technological solution to poverty and food insecurity. I hope that with this book I can deconstruct the narrative around seeds as a miracle technology. Here, I continue other social science work of "opening up the 'black box' through a critique of the dominant narrative," which "creates opportunities to open up knowledge dialogues."[46] By rejecting the Green Revolution narrative, we can open the space for a more diverse coalition of actors working to eliminate hunger and poverty.

Unfortunately, research institutions feel that they must defend the Green Revolution narrative to justify continued funding for public agricultural research. Prior to and after the Green Revolution, foundations and governments came together to establish several international agricultural research centers, each focused on certain staple crops. Today these centers are part of the CGIAR network and include CIMMYT, the research center that evolved from the RF's Mexico project, which focuses on wheat and maize.[47] The purpose of these research centers is to provide public goods and services such as monitoring diseases, collecting and providing plant-breeding materials to countries, and coordinating public sector research. Despite the many critiques of the Green Revolution, wide adaptation became embedded in some of these international agricultural research programs, particularly CIMMYT. As a result, these research organizations have struggled to address social equity. Efforts such as farmer-focused research have failed to provoke long-term institutional change.[48] In this book I investigate the historical context of wide adaptation to demonstrate its impact on current research trajectories.

In the years following the Green Revolution there have been many references to the need for a "second Green Revolution." Two programs even include this in their titles: the Alliance for a Green Revolution in Africa, sponsored by the Bill and Melinda Gates Foundation, and the government of India's "Bringing the Green Revolution in Eastern India" campaign. Organizations like the CGIAR, the Gates Foundation, and the US Agency for International Development (USAID) all take a very research- and technology-focused approach to agricultural development. These efforts apply a top-down, seed-centric approach to technology development and dissemination. This is evident in their strong focus on plant breeding, biotechnology, and other technological approaches to agricultural development (as opposed to low-technology approaches like improving farmer education or supporting infrastructure improvements). These organizations have realized, however, that certain aspects of the Green Revolution are no longer socially palatable, such as the focus on wealthier farmers. Bill Gates has stated that "the next Green Revolution must be guided by small-holder farmers, adapted to local circumstances, and sustainable for the economy and the environment."[49] Unfortunately, Gates and others do not challenge the foundations of the Green Revolution narrative. I and others argue that these organizations will fail to achieve their stated goals of reducing poverty and food insecurity through a technological approach that glosses over the context-specific nature of agriculture.[50]

Recent scholarship examines narratives not just for their pervasiveness in modern agricultural development but also for how they are used to limit discussions around alternative innovation pathways.[51] There is a tendency "for powerful actors and institutions to 'close down' around particular framings, committing to particular pathways that emphasize maintaining stability and control."[52] Pielke and Linnér emphasized that the Green Revolution narrative can also "distort thinking and action related to innovation policies more generally, if scientists and policymakers attempt to 'recreate' the supposed dynamics underpinning the first Green Revolution."[53] Maintaining the Green Revolution narrative has been an important, though implicit, role of the CGIAR to justify its relevance. Unfortunately, this limits how we imagine innovation for food security and livelihood improvement. Futures are framed around a flawed understanding of the past.

Narratives of the Green Revolution are obviously important to current agricultural research for development. But how successful are challenges to the dominant narrative? Unfortunately, not very. The Green Revolution style of research has proven resilient to decades of criticism. To help explain this incredible resilience, I turn to the concept of technological momentum as described by Thomas P. Hughes.[54] Hughes developed this concept to explain the evolution of large-scale technological systems and the persistence of certain features. These systems develop under certain political and social influences, such as the desire to centralize control over a system. Over time these systems acquire momentum of their own as a result of the buildup of physical infrastructure that reflects those initial conditions. According to Hughes, this momentum makes older systems more immune to outside forces. Technological momentum helps explain why older, large-scale sociotechnical systems appear to act in technologically deterministic ways.

Hughes's theory demonstrates why the initial conditions of research systems are influential in the trajectory of current programs. As I show in chapters 1 and 2 of this book, both CIMMYT and India's modern wheat research systems focused on breeding wheat for high-fertility conditions. As the narrative of the Green Revolution solidified, these wheat research programs continued along these same pathways: they focused on improving wheat through a centralized breeding system and then testing post hoc for adaptation. If we can view CIMMYT and the Indian wheat research system as large-scale technical systems (I believe we can), technological momentum helps explain why, despite so many calls for reform, these systems are resistant to change.

Hughes wrote that as systems mature, their associated physical infrastructure locks in the technological trajectory.[55] This is also true in agricultural research. CIMMYT's and other international agricultural research centers' strong historical reliance on plant breeding, for example, creates technological momentum as well as institutional path dependency.[56] Kenneth Dahlberg wrote that "the momentum of early staffing patterns—where plant breeding was predominant—has continued up to present."[57] The historical innovation pathway of plant breeding created a system of seed banks and research labs that profoundly influences future conceptions of technologies for climate change adaptation. The organization of scientific training, funding mechanisms, and research infrastructure of agricultural systems is largely focused on producing plant varieties. Thus, it is not surprising that plant breeding has remained a central aspect of agricultural research for development.

Actors who benefit from the dominant technological system also uphold it through social pressure. Prominent administrators and researchers in CIMMYT and India's wheat research programs, who are usually plant breeders, have maintained the social importance of wide adaptation and Borlaug's legacy. Although Borlaug never lived in India, his presence is strong through programs like the Borlaug Institute for South Asia, a collaboration between CIMMYT and the Indian Council for Agricultural Research. CIMMYT also commissioned a statue of Borlaug to overlook India's National Agricultural Science Complex grounds in 2013 (where I was based during my research in India; it was installed just a few months after I left). Researchers who don't adhere to the dominant system are often marginalized, although the culture is starting to shift at CIMMYT.

These themes are central to how CIMMYT and the Indian wheat research system have both succeeded and failed over time to achieve their stated goals. Unfortunately, these systems are not very flexible to shift focus toward poverty reduction and food security of rural people. To achieve these goals, both the scientific and donor community must move beyond simplistic technological solutions and embrace the messy reality of agriculture.

AMERICAN AGRICULTURAL ASSISTANCE IN THE MID-TWENTIETH CENTURY

Agricultural research profoundly transformed during the twentieth century. The rediscovery of Mendelian genetics, the invention of hybrid maize, and the advent of synthetic fertilizers all restructured the very business of agricultural research. These technological achievements fueled the idea that so-called developing countries could be politically transformed by technological progress. This is a key tenet of modernization theory, an ideology held by many RF administrators as well as politicians and policy advisors in the 1960s. Borlaug's work complemented the United States' political goal of pacifying nonaligned countries with food, under the presumption that well-fed populations did not revolt. As such, Borlaug's wheat became a vehicle for American values around food production.

Modernization theory and other midcentury sociopolitical contexts are important to understand how Borlaug and other American plant breeders became de facto diplomats. The esteem in which US policymakers and advisors held these scientists, and the many ties between the US foreign policy community and the RF, are also critical to explain this era of plant scientists as state builders. Globally, agricultural development shifted from an imperial to a geopolitical regime as colonialism declined after World War II. Simultaneously, the RF and other foundations rapidly proposed agricultural modernization schemes for formerly colonized countries. These agricultural development schemes assumed that other countries needed to "develop" along a "social evolutionist teleology."[58] Modernization theorists held that history could be "sped up" and that societies could become modern given the right set of knowledge and inputs.[59] Importantly, modernization theory heavily relied on technology as a tool of development.[60]

Agricultural modernism began after World War II, although the roots of modernism can be traced much further back.[61] In the immediate post–World War II era, the United States entered international development through policies such as the Marshall Plan to help rebuild postwar Europe. The United States now involved itself in international food politics as a form of development, against previous isolationist policies. US politicians strategically leveraged surplus grain through policies such as Public Law 480 (the Agricultural Trade Development and Assistance Act, or PL-480) in 1954, which provided food aid to other countries. Prior to this, food and agriculture were not part of US international policy, and aid was given as needed.[62] After World War II, food became a keystone of international policy.

Simultaneous to the United States' involvement in international food assistance, fears of the population problem escalated in the 1940s and into the 1960s as a foreign policy concern. New tools such as population models led to dire predictions of overpopulation and starvation.[63] Increasing the food supply became a popular solution to the "population problem." US foreign policy experts framed the problem as a "gap, in certain areas, between the rate of increase in population and the rate of increase in local food production, yielding a slide into increased dependence on U.S. food surpluses."[64] Policymakers wanted developing countries to become self-sufficient in grain production, and thus engaged in support of agricultural development schemes.

India, for example, became a major site of PL-480 intervention and agricultural development. India became independent in 1947, after experiencing a devastating famine in West Bengal in 1943. In the 1950s the US government began food and technical assistance programs in India, and various foundations also became involved in India's agricultural development. Then in 1965 and 1966, India was in an ecological and political crisis: a war with Pakistan, a drought, and the untimely deaths of successive prime ministers Jawaharlal Nehru and Lal Bahadur Shastri.[65] US president Lyndon Johnson utilized this tumultuous time to implement a "short tether" approach to PL-480 food aid: withholding aid to negotiate several agricultural policy changes, including seed and fertilizer imports, a minimum support price for wheat, and ongoing land policy reforms such as abolishing absentee landlords.[66] At this same time, the RF helped restructure the Indian agricultural research system and introduced new wheat varieties into Indian fields.

Why was Johnson so personally invested in Indian agriculture? His investment was in part owing to his personal vision of agrarian "self-help" and the Great Society, but also to a desire to sway India away from communism and toward democracy.[67] As John Perkins explained, Cold War intellectuals made the link between population, hunger, and national security, theorizing that overpopulation and hunger would cause political instability in the developing world.[68] Newly independent states, as well as other "developing" countries, became known as the "Third World" in the 1950s to represent their potential mobilization as well as their nonaligned status in the Cold War.[69] The United States could wage a war on hunger as a politically palatable Cold War intervention.

Organizations such as the RF and the Ford Foundation, USAID, and the Food and Agriculture Organization were well positioned to offer solutions in the war on hunger. The RF had been supporting agricultural research since the early 1900s.[70] In the 1940s, the RF became involved in agricultural research in Mexico, a site of both political and economic importance to the United States. During World War II President Franklin Roosevelt focused on strengthening economic ties in the Americas to resist communist and fascist powers.[71] At the same time, the RF had to end its programs in Europe, and it saw moving to Latin America as a natural adaptation of these programs. The RF saw Mexico as a surrogate for other developing countries.[72]

THE ROCKEFELLER FOUNDATION'S INTERNATIONAL AGRICULTURAL PROGRAMS, 1943–1966

The RF's program in Mexico established agricultural research as a mode of international agricultural development. In 1940 President Roosevelt sent Vice-President Henry A. Wallace to Mexico to investigate the problems of agriculture there. Wallace returned from the trip suggesting that the RF work on improving maize production. The next year the RF sent a group of scientific advisors to Mexico to survey the possibilities for an agricultural program in the country. Elvin Stakman, Richard Bradfield, and Paul Mangelsdorf, all professors of agricultural science, went to Mexico and developed a set of recommendations for a technical assistance program.[73] Their recommendations included improving maize, wheat, and bean varieties, and promoting better agronomic management practices, ostensibly under the leadership of American scientists.[74]

The RF entered an agreement with the government of Mexico in 1943 that would be known as the RF's Mexican Agricultural Program (MAP) under the Oficina de Estudios Especiales (Office of Special Studies). The MAP was hosted by the Escuela Nacional de Agricultura de Chapingo (National College of Agriculture of Chapingo), outside of Mexico City.[75] The MAP started under the scientific leadership of J. George Harrar and initially focused on improving maize, beans, and wheat for Mexican conditions. The RF aimed to bring agricultural expertise to problems facing Mexican agriculture, such as breeding varieties of wheat that were resistant to stem rust and developing improved varieties of maize.[76] Borlaug was hired in 1944 to work on the wheat program. Borlaug and

Harrar had both done their doctoral work with Stakman at Minnesota.[77] Many of the important details of the MAP are well documented in the secondary literature; therefore, the next sections will focus on elements of the MAP that are relevant to Borlaug's international wheat program and how this became a model for international agricultural research.

During its early years, three principles of the MAP emerged that became staples of later RF programs: collection of plant genetic materials, multilocation testing of RF-developed crop varieties, and a focus on high-yielding varieties.[78] Each of these was exemplified in Borlaug's work, but of importance here is the focus on crop yield. Warren Weaver, director of the RF's natural sciences division, was influential in framing the work of the RF as promoting more food per area. He proposed increasing crop yields as a solution to population growth, which would neatly solve concerns about rapidly growing, urbanized populations causing the erosion of surrounding agricultural lands.[79] As the ability to establish new cropland declined, RF scientists framed increased fertilizer use and higher-yielding crops as a solution to this problem. This was critical to Borlaug's wheat program and future RF international programs.

Around 1950 RF administrators began considering the long-term future of the agricultural program in Mexico. Minutes from the 1950 meeting of the RF's International Health Division state, "Five years ago we got into agricultural development in Mexico. . . . *The success and interest in it would in itself raise the question whether The Rockefeller Foundation should, as a general undertaking, continue to do that kind of program.*"[80] While some RF employees disagreed with the "universal approach" of the MAP, by the end of the decade it was clear that the RF would proceed with country-specific and transnational (cooperative) agricultural programs based on the MAP. In other words, they would continue hiring foreign agricultural scientists to improve yields of staple crops of specific countries and regions.[81]

In 1950 Harrar left the MAP to become deputy director of agriculture for the RF in New York. Harrar was particularly important in expanding the MAP model to other countries in Latin America and Asia.[82] Starting in 1950, the RF began a series of country-specific agricultural programs that were based on the MAP and focused on maize and wheat, but also included livestock and other commodities, with MAP-like programs in Colombia (1950), Chile (1955), and India (1956). Simultaneously, the RF started several cooperative programs. These programs aimed to facilitate the international spread of improved plant genetic material (called germplasm, since not all plants are propagated by seed); and assist in building and training scientific researchers and staff in developing countries without stationing RF staff directly in those countries.[83] Throughout the 1950s and 1960s, the RF's cooperative agricultural programs grew rapidly. The foundation developed a network of cooperative crop testing programs through South America, the Middle East, and Asia, as a means for the RF to institutionalize their research agenda far beyond Mexico.[84] These cooperative programs functioned as an internationally diffused MAP.

In the early 1950s several central American countries came together to request RF assistance with maize production.[85] This led to the RF's first international cooperative program, the Central American Corn Improvement Program, which started in 1954 and was based in Mexico and Colombia.[86] The program's goal was to test maize varieties that had already been exchanged between Colombia and Mexico to "see whether some of them may be used at once in the cooperating countries."[87] Two years later the RF started a worldwide maize testing program that extended to several more countries in South America and India, Indonesia, and the Philippines.[88] The worldwide program aimed to evaluate the "adaptability and genetic value of specific material throughout the world, help breeders learn what is available, and help the germ plasm banks to fill seed requests intelligently."[89] This program both served the needs of cooperating countries and advanced the RF's agenda of distributing the foundation's agricultural methods and crop materials. The maize program initially focused on maize but later included wheat and potatoes. This morphed into the Inter-American Food Crop Improvement Program in 1959, led by the RF's Edwin J. Wellhausen.[90]

A similar process led to a cooperative program for wheat. Wheat scientists agreed to establish an inter-American cooperative yield test for wheat at the Fourth Latin-American Conference of Agricultural Scientists, held in 1958. RF scientists, specifically Borlaug, would coordinate this program from Mexico. In the late 1950s, RF administrators were eager to get Borlaug into a position of international leadership for wheat science. Harrar wrote to Borlaug in 1958 to say that "it is now timely to begin to intensify international research on small grain improvement in the Americas and its logical leadership to this effort should come out of the cooperative agricultural program in Mexico."[91] Harrar also wrote in 1959 to José Vallega of the Food and Agriculture Organization that "we now want Dr. Borlaug to operate on a very much more international scale. We would like to support him in an effort to strengthen cereals improvement research throughout the Americas and link these more closely together from the northern to the southern extremes of production areas."[92]

While Harrar and others were supporting Borlaug's international leadership, a new iteration of the MAP was necessary. The MAP was never intended to be a permanent project in Mexico. Starting around 1960 it moved toward a complete administrative transfer to Mexican scientists, many of whom the RF had trained. In 1961, the administrative portion of the MAP was terminated, and the Instituto Nacional de Investigaciones Agrícolas (INIA, National Institute of Agricultural Investigations) was formed to take over the MAP's national operations in Mexico.[93] Many of the RF scientists were unhappy working at the INIA due to budget constraints and political tensions, so, in 1963, the RF partnered with the Ford Foundation and government of Mexico to form the International Center for Corn and Wheat Improvement, headquartered in Chapingo, Mexico, and directed by Wellhausen. The RF still provided funding to the INIA but built a new scientific complex to house the RF researchers.[94] The International Center for Corn and Wheat Improvement allowed the RF to continue to operate internationally from Mexico rather than pulling operations and scientists out of the country.

The conceptual shift to international research programs was a deliberate move by RF administrators such as Albert Moseman, who had spent time abroad with the US Department of Agriculture before joining the RF, and Dean Rusk, who came from the US Department of State. The International Center for Corn and Wheat Improvement (the Center), as its name suggests, focused on international research programs for maize and wheat. These programs built on the existing transnational infrastructure in the Americas and expanded over the next few years to include collaborators in Africa and Asia. The Center's overall goal was "to aid, on an international scale, in the improvement of materials and methods for the production of maize and wheat by obtaining improved varieties and by applying breeding techniques to achieve greater protection against insect pests and diseases as well as destructive climatic effects," as described by the secretary of agriculture and livestock of Mexico, Julián Rodriguez Adame.[95] The main goals of the wheat program included developing new varieties of wheat that were rust resistant and "high-yield, widely-adapted."[96] For maize, the goals were to collect and distribute germplasm, to breed varieties resistant to disease, to develop varieties for high-fertility conditions, and "to develop corn varieties insensitive to day length and temperature, thereby increasing adaptability."[97] Wide adaptation was a pillar of the RF's international programs, theoretically allowing researchers in one location to produce technologies with broad applicability.

Not surprisingly, this was exactly the time that Borlaug and others began to popularize wide adaptation as a breeding goal for cereal crops. However, RF scientists were finding that maize was not as successful abroad. It had a narrower range of adaptation, likely because of its sensitivity to day length and other environmental sensitivities, genetic or otherwise. By 1965 it was increasingly clear that wheat would be the main international focus of the Center, thanks to its ability to grow well under a variety of conditions and locations. Yet maize research was still, and remains today, an important component of the Center and its later evolutions.

Plant breeders formed the core staff from MAP's beginning and were vital to the RF's agricultural programs in later periods. Lewis M. Roberts, an associate director of agricultural sciences at the RF, wrote in a 1965 report that the Center's asset was their four wheat breeders: Borlaug, R. Glenn Anderson (who had been recently hired by the RF to work in India), John W. Gibler, and Charles F. Krull, who were both recently transferred from Colombia to the Center's headquarters in Mexico. Roberts viewed the location of Mexico as an asset as well. He wrote, "the broad range of ecological conditions in that country provide a highly favorable natural setting for maize and wheat improvement work applicable to a broad belt of the globe, especially in the tropical latitudes."[98] This was central to the theory of wide adaptation: that varieties developed under one set of conditions could nonetheless thrive in a wide range of environments.

The International Center for Corn and Wheat Improvement became CIMMYT in 1966. CIMMYT was governed by an international board of RF and Ford Foundation affiliates as well as international scientists that participated in CIMMYT's international programs.[99] Wellhausen became CIMMYT's director general and Borlaug led the wheat program. This change also meant that CIMMYT was more autonomous from the RF, though clearly still financially dependent. The RF continued to sponsor the country-specific programs, such as India. All of the RF country-specific, coordinated, and international programs shared a common theme: reducing agroecological complexity down to a uniform prescription of fertilizers, irrigation, and high-yield-potential varieties. This is reflected in the crop-specific nature of many of the programs, and the assumptions that technologies could scale across large agroclimatic zones. Wide adaptation became central to CIMMYT's philosophy in the 1960s and remains so today.

In this book I examine how wheat research went from a location-specific activity to a centralized program both internationally, in the case of CIMMYT, and in India. My focus on India is due not only to the availability of archival information in India but also to the landscape of post-Independence India and the involvement of RF scientists, particularly Borlaug's interest and participation in reshaping Indian wheat science. In chapter 1 I explore Borlaug's program on widely adapted wheat in the 1960s, the link between Borlaug's international wheat trials and his ideas about wide adaptation, and the accounts of his collaborators, Charles Krull and Keith Finlay. In this chapter I set up the exciting time of Borlaug realizing that his Mexican-derived semidwarf wheat varieties were outyielding local varieties from South America to the Middle East and South Asia. We see a bit of conflict as Borlaug and Krull's colleagues in the Food and Agriculture Organization argued for testing under local conditions. But the real drama starts in chapter 2, as Borlaug rapidly pushed for the release of widely adapted varieties in India in the mid-1960s, meanwhile helping reshape the structure of Indian wheat research in a way that strongly aligned with his program on wide adaptation. In chapter 3 I show the state of Indian wheat research in the immediate aftermath of the Green Revolution, and how Indian scientists renegotiated ideas around wide adaptation. Chapter 4 skips forward to the present time and includes excerpts from my interviews with Indian wheat scientists, the legacy of wide adaptation, and an assessment of Indian agricultural innovation systems. In chapter 4 we see that Borlaug's ideas about wide adaptation and testing under ideal conditions are embedded in the policies and bureaucracy of Indian wheat research, despite criticisms and attempts at reform since the Green Revolution. I show how present-day wheat scientists struggle to move past the dogmas of wide adaptation. Finally, in chapter 5 I examine three case studies of failed wide adaptation: first I return to the Mexican Agricultural Program and a maize program designed to benefit smallholder farmers, then I explore post–Green Revolution wheat programs in Turkey, North Africa, and the Middle East. In these cases, RF- and CIMMYT-affiliated researchers were responding to pressure to recreate Borlaug's success with spring wheat and to criticisms that the Green Revolution created wealth inequities. And in all cases, wide adaptation failed to solve the local and regional problems of food production.

Wide adaptation has never been unproblematic. Wide adaptation and the concepts around it have been some of the most discussed and debated issues in agricultural science. Yet most historical accounts of the Green Revolution either mention wide adaptation in passing or not at all. It has become so embedded in certain types of wheat research that it is dogmatic (though still contested by outsiders). The problem is that wide adaptation is a specific mode of research that has many major drawbacks in terms of social justice. The philosophy of wide adaptation has caused neglect of the problems of rainfed and dryland agriculture, where smallholder farmers are predominantly located. It promises varieties that are plastic to environmental change, but there are physiological limits to the plasticity of modern wheat varieties to stressors such as heat and drought. The promise of wide adaptation has also been used to justify the dominance of plant-breeding solutions to complex problems such as food security. The world may very well need more agricultural extension workers than plant breeders, but donors view agricultural extension, or education, as a location-specific activity, while plant breeding—because of wide adaptation—is viewed as an international activity producing public goods.

One of the more positive outcomes of the Green Revolution was a shift from viewing farmers as backward peasants to entrepreneurial agents. But the philosophy of wide adaptation ultimately preserves the divide between scientists as innovators and farmers as passive receivers. It heralds universal solutions to problems such as food insecurity, resource scarcity, and climate change, while ignoring the complex sociopolitical as well as biophysical realities behind food production and food security. Ultimately, those engaged in agricultural research development must recognize that wide adaptation has not borne the fruits it promised. Alternative strategies are necessary to address the multidimensional challenges of small farmers, rural populations, and food-insecure people around the world.

Agricultural development has always been more political than its practitioners want to admit. It's tempting to say, "We're only transferring technology," and to ignore the ways that technology can exacerbate social inequities or constrain so-called beneficiaries to a certain path. My appeal to those working in agricultural research and development today is to consider the social consequences of developing and promoting new technologies, which will likely require working with social scientists and humanists. While technology seems a promising way to bypass the messiness of politics and social change, technologies are seldom neutral and can reinforce imperialist and exploitative modes of development.

CHAPTER I

NARRATIVES AROUND WIDE ADAPTATION IN INTERNATIONAL WHEAT RESEARCH, 1960–1970

NORMAN E. BORLAUG, CHARLES F. KRULL, AND KEITH W. FINLAY

In this book I trace Norman Borlaug's controversial ideas around adaptation from the 1950s and through several decades and countries. As such, a bit of background on the concept of adaptation is necessary. In the ecological and evolutionary sense, adaptation is a heritable process that contributes to a species' survival and fitness in its environment. The field of evolutionary biology considers adaptation a process, but in agricultural science, adaptation is more of a state or condition.[1] Historian Emily Pawley explained how the term *adaptation* sometimes describes human intervention into biological systems, such as a farmer *adapting* livestock to a particular environment through breeding and acclimatization; this concept is still present in the agricultural sciences.[2]

The adaptation of a plant includes its physiological tolerance and requirements of temperature, soil composition, moisture, disease, sunlight, wind, species competition, and so on. Through evolution and natural selection, as well as artificial selection (by farmers and plant breeders), it is commonly assumed that agricultural plants are specifically adapted to their region of origin—the place that they evolved in. Since at least the mid-1800s agriculturalists have used the term "wide adaptation" to describe the agroclimatic range of horticultural species in the United States—for example, the *Report of the Commissioner of Patents for the Year 1853* includes the observation that "the principal species [of the tobacco plant], *Nicotiana tabacum*, is sufficiently variable and sufficiently capable of a wide adaptation to permit Cuban varieties to be immediately transferred to Ohio or New York."[3] Adaptation is one of many possible plant characteristics, albeit a helpful one in terms of agricultural development. In the early twentieth century, wide adaptation was not often pursued as a strategy, however, as most plant breeders focused on developing crops for specific locations.[4]

In a plant-breeding context, adaptation means the relative performance (roughly, the yield and disease resistance) of a plant variety under different conditions. A widely, or broadly, adapted variety gives high yields under many different environments and locations. Wide adaptation can also be defined as phenotypic stability plus high yields. Specific, or narrow, adaptation refers to a variety that thrives only under a specific set of environmental conditions.[5] Scientists can measure how plant characteristics (such as plant height) vary to study the phenotypic responses of plants to different conditions, but adaptation is typically measured in yield (grain weight per area).

Wide adaptation existed in the lexicon of agricultural scientists in the 1960s, but only in the margins of agricultural science. The conventional wisdom of plant breeding in the early twentieth century was that crop selection should occur in the target environment, creating varieties with specific adaptation to the local conditions. Even a 1954 annual report from the Rockefeller Foundation (RF)'s Colombian Agricultural Program stated, "It is axiomatic in agricultural research that an improved crop variety, to be commercially successful in a given region, must be developed and tested in that region."[6] In other words, agriculture was a "site-specific science," and most cereal breeders viewed wide adaptation with little more than skepticism.[7]

Borlaug's wheat program changed the paradigm of international agricultural research. Borlaug introduced the promise of intentionally designing a crop that could be easily transposed between locations. For the US foundations that wanted to make their mark abroad through agricultural assistance, this was a huge boon. These foundations set up international research centers that followed Borlaug's research model and trained international scientists in his methods. Throughout the 1960s, the RF-sponsored international centers focused on developing a few widely adapted varieties of wheat, rice, and maize that could be grown in many countries.[8] While each program has continued up to the present, wheat most successfully proved itself as widely adapted.

Over the years wide adaptation has been "blackboxed": it has been packed with multiple, unfounded meanings and is only occasionally critically reviewed. In this chapter I unpack the black box of wide adaptation, starting with its role in Borlaug's wheat program. Through a series of incidental connections and rediscoveries in the 1950s and 1960s, Borlaug found that spring wheat varieties derived from Colombian and Mexican varieties had consistently high yields in widely dispersed trials. At that time, most agricultural scientists were skeptical that one variety could have consistent high performance over a variety of locations. Borlaug's international trials showed that a widely adapted variety could even outyield popular national varieties in their home countries. In just a few years, Borlaug took an unpopular idea and completely changed the paradigm of plant adaptation.

In this chapter I reveal the history of one of the most influential yet underexplored ideas in agricultural science. I explore how Borlaug came to focus his research program on wide adaptation and fertilizers; the work of his colleague, Charles Krull, in promoting wide adaptation and fertilizers; and Borlaug's correspondence with Keith Finlay, who supported Borlaug's mission but questioned his methods. We see the evolution of Borlaug's philosophy and program on wide adaptation and its acceptance in the international community as a valid scientific paradigm.

BORLAUG AND THE GLOBALIZATION OF SPRING WHEAT RESEARCH, 1950–1968

Borlaug found that wheat varieties from the United States and Canada were generally poorly adapted to Mexican conditions due to different lengths of daylight and seasons. The United States and Canada are major wheat-growing countries, but they grow winter wheat, which requires a period of cold to mature. For Mexican environments, he needed to use spring wheats in his plant-breeding experiments. Spring wheats are grown in tropical and subtropical areas and do not require a cold period. Under Borlaug's supervision, the Mexican Agricultural Program (MAP) released disease-resistant spring wheat varieties that were adapted to Mexican conditions in 1948, and by 1957 these new varieties constituted 90 percent of Mexican wheat acreage.[9]

Borlaug became interested in the idea of wide adaptation after participating in the US Department of Agriculture's (USDA) International Wheat Rust Nursery, which started in 1950.[10] In response to an epidemic of wheat stem rust in North America, the USDA set up the Wheat Rust Nursery to test their large collection of wheat seeds in different environments around North and Central America and to identify rust-resistant varieties.[11] By 1952 the nursery had expanded to Australia and various countries in Africa and Europe.[12] This nursery was possibly the first systematic global wheat test, and Borlaug was involved from its beginning.

Borlaug's mentor, Elvin Stakman, had written to the RF's president,

Dean Rusk, back in 1953 that it would be useful to breed for "the best possible combination of genes for yielding ability, disease resistance, or any other universally useful character, without considering adaptability to particular areas."[13] These were not novel ideas in agricultural science, but Stakman's idea that "these lines could then be given to breeders in all interested countries for use in developing varieties adapted to their conditions" was prophetic of Borlaug's wheat program.[14] Stakman's ideas contrast with those of his colleague J. George Harrar, who as president of the RF in 1961, stated, "Unfortunately, most scientific advances most directly benefit the particular geographic area in which they originated. This is especially true in the agricultural sciences."[15] Borlaug's work on wheat proved this false.

Borlaug developed an interest in collecting basic data on the adaptation of wheat varieties after seeing how well some of the RF's wheat varieties, such as Lerma Rojo and Nariño 59, performed in the USDA nursery.[16] By 1959 Borlaug became convinced that wheat crosses between certain foreign strains produced varieties that could be grown over wide geographic areas. He stated at a 1960 meeting that "wheat is very different from corn in that it appears to be much more flexible in its adaptation to different soils and climatic conditions."[17] Borlaug's finding contradicted what many scientists presumed at that time, which was that agricultural assistance programs would always be constrained by geography.

Around 1959 Borlaug proposed a new international wheat nursery that would prove wheat's adaptation to diverse geographies. He wrote in a trip report, "In the past there has been a great deal of circumstantial evidence that certain types of wheat have great flexibility and adaptation; however, this has never been checked experimentally, and it seems that the time has now arrived for doing so."[18] He soon proposed a "uniform yield nursery" to collect "valuable information on varietal adaptation" in wheat.[19] In 1960 Borlaug started his first international nursery, called the Cooperative Inter-American Spring Wheat Test. Borlaug sent packets of twenty-four spring wheat varieties from the Americas and Australia to twenty different locations in the Americas, as well as in Egypt, Kenya, and Pakistan, where he had collaborators and former students.[20] In the first year of trials, the RF Colombian variety Nariño 59 had the highest average yield at the eighteen reporting locations, though it ranked first in only three of the trials.[21] This was surprising because it usually takes several years to adapt a foreign variety by crossing it with local varieties.

The Food and Agriculture Organization (FAO) soon invited Borlaug to tour the Middle East, where it had been working on wheat since 1952. In 1960 Borlaug examined some of the problems of wheat cultivation in that region.[22] Derek Byerlee has remarked on the importance of this

TABLE 1.1. Countries that participated in the First International Spring Wheat Yield Nursery and number of tests, 1964–1965

Country	Number of sites
Argentina	4
Australia	2
Chile	1
Colombia	1
Cyprus	1
Ecuador	1
Ethiopia	1
Guatemala	1
India	3
Iran	1
Iraq	1
Jordan	1
Lebanon	1
Libya	1
Mexico	2
Pakistan	3
Romania	1
Saudi Arabia	1
South Africa	1
Sudan	2
Syria	1
Turkey	1
United States	2

two-month journey, which was Borlaug's first trans-Atlantic tour.[23] Borlaug visited twelve countries, including Pakistan and India, and wrote an unusually long 198-page report on the trip.[24] While traveling, Borlaug observed varieties from the Rockefeller Foundation agriculture program planted in the nurseries and other experiments and was "amazed to see the wide adaptability of many of the wheat materials" from Mexico and Colombia.[25] He felt that the scientists running the nurseries did not recognize this amazing feat for what it was, owing to their lack of experience outside their own country. Based on these initial results, Borlaug wanted to expand his own wheat yield trials to the Middle East, India, and Australia.[26] He planned to use the international trials to evaluate "the relative adaptability of a uniform set of varieties of different origins by growing and observing them systematically under widely different conditions of climate, soil, and latitude" as well as the "possibility of developing wheat varieties with extremely wide patterns of adaptation."[27]

The RF and FAO together started the Cooperative Near East–American Spring Wheat Yield Nursery in 1962. Borlaug again packed seeds from twenty-five varieties of spring wheat into hundreds of envelopes, including commercial varieties from the Middle East, two varieties from Colombia, and seven varieties from Mexico. All varieties were grown under widely varied conditions, as Borlaug recommended planting seeds on uniform plots that represented average local conditions. In the first two years of trials, five Mexican varieties yielded, on average, the highest of all twenty-five varieties entered in the trials.[28] These varieties were among the highest yields even under unfertilized and rainfed conditions.

In 1964 Borlaug combined the Inter-American and Near East–American nurseries into the International Spring Wheat Yield Nursery. He sent 25 varieties to 34 locations in 23 wheat-growing countries (see table 1.1). Like the previous nursery, seeds were grown under both irrigated and rainfed, and fertilized and nonfertilized conditions.[29] And again, five Mexican varieties yielded the highest, on average. Draft RF reports casually noted the wide adaptation of the Mexican varieties, but as time went on the RF researchers made a stronger case that wide adaptation was not just achievable but desirable.

Bolstered by the results of his international trials, Borlaug spent little time pondering the theoretical aspects of wide adaptation and quickly moved to implementation. Borlaug and his colleagues saw wide adaptation as a method to share wheat varieties with countries with limited scientific resources.[30] He wrote to RF agricultural sciences director Albert Moseman in 1963 that materials from "one broadly based wheat breeding program" focused on wide adaptation can be "reselected for direct use in countries far distant from the location of the breeding programs."[31] This could radically speed up the time it would take to adapt varieties to a new location through crossbreeding. In 1965 Borlaug made a case for the moral imperative of wide adaptation, writing that "varieties and breeding lines with broad adaptation can be introduced rapidly and grown successfully in many areas of the world where expansion of food production is urgently needed. This is not possible with narrowly adapted varieties."[32] Borlaug realized that he could not only transmit scientific knowledge to other wheat breeding programs around the world but also directly transfer wheat seeds.

Borlaug's theoretical explanations for the wide adaptation of Mexican- and Colombian-derived wheats evolved over the years. Borlaug initially recognized that wide adaptation was the result of certain "germ plasm complexes" that were genetically inherited.[33] He surmised this because varieties he derived from the lines Mentana (from Italy), Marroqui/Florence-Aurore (from Tunisia), and Gabo (from Australia) tended to be more adaptable across locations.[34] Borlaug later attributed wide adaptation to his unique method of wheat breeding. Around 1945 Borlaug began growing wheat generations alternately between north and central Mexico to speed up the time needed to select and stabilize a new variety, which typically takes about ten years.[35] This was later called "shuttle breeding," and is one of Borlaug's best-known legacies.[36] In the winter, Borlaug planted wheat in the Sonora region of Mexico—a coastal, irrigated region near sea level and at 28°N latitude. Then he would select the best offspring from that season and plant them in Toluca (near Mexico City), which was at 18°N latitude and had a high altitude, heavy rainfall, and a higher prevalence of pathogens. Borlaug insisted that shuttle breeding would produce results. He stated in his 1967 oral history: "We were constantly, and very early, we were doing it consciously—discarding those things that fit in only one environment. We were interested because of the ease of multiplication of varieties of having things that were broadly adapted and consequently probably less vulnerable to the vagaries of climate, but also that if we found a variety that was well adapted and yielded well—it could be grown widely in Mexico."[37] Borlaug's insistence that wide adaptation was purposeful conflicts with other sources and my interviews with scientists who knew him, which described the finding as serendipitous.[38] Borlaug retroactively credited his shuttle breeding experiments with providing the proper selection pressures to favor widely adapted varieties.

Within a few years, however, Borlaug realized that the main genetic contributor to wide adaptation was photoperiod insensitivity, meaning a crop that is not sensitive to day length. Wheats from the United States and Canada were photoperiod sensitive, while photoperiod-insensitive wheats could be grown in a variety of latitudes, elevations, and seasons. Borlaug wrote that "in all probability one of the important factors in this lack of flexibility is their sensitivity to change in day length and date of planting."[39] Photoperiodism was discovered in 1918 by USDA researchers W. W. Garner and H. A. Allard, so by Borlaug's time it was well known.[40] Borlaug hypothesized that his shuttle breeding method had resulted in selection that favored photoperiod-insensitive varieties that thrived in both the Sonora and Toluca regions, which have different seasons and photoperiods.[41] Borlaug later wrote that due to the "day-length insensitivity and broad-based rust resistance" and high yields of the Mexican semidwarf wheats, countries could release "only a few varieties needed to serve commercial farmers—rather than a dozen or more that would have been necessary if narrowly adapted varieties would have been developed."[42] This would simplify "the work of newly formed national seed agencies."[43] Borlaug was correct that day-length insensitivity and rust resistance allowed countries to adapt foreign varieties to their conditions much more rapidly than in the past, because they did not require crossbreeding with local varieties. It should be noted, however, that the photoperiod insensitivity is not possible in all crops.

Although Borlaug clearly recognized photoperiod insensitivity as the main component of wide adaptation, his research program moved toward developing "even more widely adapted genetic types" of wheat and asked, "What is the maximum range of adaptation that can be incorporated into a variety?"[44] Borlaug seems to have thought that there were additional genetic factors of wide adaptation besides photoperiod insensitivity. And indeed, there was: Borlaug's varieties were bred to withstand high levels of fertilizer.

When Borlaug started working for the RF in Mexico in the 1940s, his task was to develop wheat varieties that had higher yields and greater disease resistance than the local varieties. He realized that more nitrogen-based fertilizer was required to improve yields. But when too much fertilizer was added to local wheat varieties, they would fall over because of the heavier grain at the end of the tall, thin stalks. This is called *lodging*, and it can also be caused by high winds or rain. A solution to the problem of lodging appeared when Borlaug learned about "dwarf" wheat through Orville Vogel, at Washington State University. Vogel had obtained the dwarf variety Norin 10 from Japan. Dwarf and semidwarf wheats have shorter and thicker stalks than traditional wheat varieties. Semidwarf wheats can withstand higher levels of fertilizers without lodging, which means semidwarfs typically have a higher yield potential than the traditional tall wheats. Borlaug began crossing Norin 10 with Mexican wheat varieties in the 1950s, which resulted in a semidwarf wheat variety adapted to Mexican conditions. By 1955 Borlaug had successfully crossed Norin 10 with Mexican varieties, and in 1962 he released the semidwarf wheats Pitic 62 and Penjamo 62 in Mexico.

Even before the semidwarf varieties, Borlaug was already adapting wheat varieties to higher-fertility conditions starting around 1945.[45] He assumed that fertilizers would soon become more easily available and affordable globally. Borlaug saw fertilizer inputs as key to reducing lost soil fertility from centuries of extractive farming. By the mid-1950s Borlaug tested new wheat varieties under high-fertility conditions exclusively. He believed that varieties must be adapted to higher-fertility conditions to increase overall food production. At Borlaug's suggestion, Argentina's varietal improvement program was "reoriented in 1962 in order to develop varieties which would be better adapted to higher levels of soil fertility should the use of chemical fertilizers become widespread."[46] Borlaug reasoned that "any breeding program which did not take into consideration a change in levels of soil fertility within the next five years, would be doomed to failure."[47]

Borlaug also believed that planting wheat under favorable environments (high fertility and optimum irrigation) allowed the scientist to observe a variety's "true genetic potential," because variation between varieties would be more obvious.[48] In a letter to a scientific advisor in West Pakistan in 1964, Borlaug argued that at high fertility levels, one can see problems with the wheat variety not evident on "tired soil."[49] Borlaug also emphasized that results from irrigated trials were more reliable than those from rainfed trials because the rainfed trials had more environmental variation that would eclipse genotypic differences.[50] He also noted that working under low-fertility conditions slowed down the plant-breeding process. He wrote in 1960 that RF scientists were "spending upwards of 70% of their time trying to unsnarl the problems relating to soil fertility, instead of devoting all or most of their efforts to the aspects relating to crop breeding and crop management."[51]

Finally, Borlaug believed that varieties adapted to higher levels of fertilizer would lead to social change among farmers and scientists and overall higher levels of wheat production. He wrote in 1966 that the government of West Pakistan "should realize that solving the fertilizer problem for wheat will be the start, not the end, of increased fertilizer demand. For once a farmer learns how to use fertilizer in large dosage on wheat, the practice will quickly spread to other crops. That was our Mexican experience."[52] In India, Borlaug argued, "the program should try to produce tremendous yield increases on the area where the dwarf varieties can be heavily fertilized and properly watered."[53] He continued, "By so doing a complete change in the psychology of wheat production— from one of survival to one of high yields—will shock both the farmer and the scientist."[54] Borlaug believed that complacency of local agricultural scientists was one of the biggest hurdles to modernizing agriculture, and that they needed a shock to wake up.

When Borlaug began focusing on wide adaptation around 1960, his wheat research program was solely focused on selection and testing under favorable conditions. Borlaug made wide adaptation a key part of his research when he became head of the RF's international wheat program. To Borlaug, wide adaptation was a symbol of his program's global reach and ability to cause radical agricultural change. Against the prevailing sentiment that "plant breeders must work in the place where their crop will be grown," Borlaug argued that wide adaptation was not only a tenable but also a desirable plant-breeding goal.[55] He influenced agricultural scientists around the world through his trainings, publications, correspondence, and lectures. Beyond this paradigm-shifting endeavor, however, Borlaug had a very mission-oriented reason to promote wide adaptation. He wanted to transform agriculture in developing countries from premodern to modern, and thought that widely adapted, fertilizer-responsive varieties were the most likely way to accomplish this.

While Borlaug was breeding and testing wheat under high levels of fertilizer, much of the developing world was relying only on natural soil fertility. For the international trials, the plant scientists in the Mexican locations applied 80 to 120 kilograms per hectare (kg/ha) of nitrogen (N) and sometimes more. A 1969 review of CIMMYT's research found that "one rate of fertilizer (160 pounds of nitrogen per acre) is used throughout the 140 acres of experimental plots devoted to wheat" (160 pounds per acre is about 179 kg/ha).[56] This rate was comparable to the highly fertilized Belgium, which between 1962 and 1966 used an average of 158 kg of N per arable hectare (including crops other than wheat).[57] India, on the other hand, barely registered at 3.3 kg N/ha (again, for all crops).[58] Pakistan, Iraq, Iran, Syria, and Turkey, as well as Africa's major wheat-producing countries, all consumed less than 10 kg N/ha during this period.[59] Despite the massive gap between fertilizer rates at CIMMYT and the collaborating countries, Borlaug soldiered on with his international wheat program.

When Borlaug started doing research in the Sonora region, the RF initially did not support him because this was outside of the program's mandate to help peasant farmers. Farmers in the Sonora were wealthier and had the benefit of irrigation, while the central Toluca region had smaller farms, poorer farmers, and more varied environmental conditions. Despite these differences, most Mexican farmers quickly adopted wheats derived from the RF program because of their high yields and disease resistance. RF-derived wheat varieties also spread fairly quickly in Colombia, Guatemala, Ecuador, Chile, and Bolivia.[60] The RF's maize program was not as successful, however, because maize was not as adaptable as wheat. The lack of fertilizers, irrigation, and government support also slowed down the spread of RF wheats in some Latin and South American countries.

Around 1965 Borlaug began promoting the idea that widely adapted varieties were adapted not only to different geographies but also across agroclimatic conditions such as irrigation and soil fertility. In response to those who might criticize his focus on favorable environments, Borlaug wrote that "even at low fertility and on dryland, they [semidwarf wheats] do surprisingly well, displaying their efficiency even though they were developed under irrigation."[61] Borlaug saw the success of his varieties in his international trials and used these results to support his claims. According to Borlaug, "because of this mass of information . . . we feel pretty confident also in moving aggressively in Pakistan and India or in Turkey."[62]

Farmers quickly adopted semidwarf, fertilizer-responsive, and photoperiod-insensitive wheat varieties in certain regions, but especially in the irrigated parts of India, Pakistan, and coastal Turkey. US Agency for International Development (USAID) administrator William Gaud declared the Green Revolution in 1968 and Borlaug was awarded the Nobel Peace Prize in 1970. In his Nobel lecture, Borlaug said that the Mexican wheat's "unusual breadth of adaption" along with other factors "made the Mexican dwarf varieties the powerful catalyst that they have become in launching the green revolution."[63] Thus, Borlaug canonized wide adaptation in his narrative of the Green Revolution.

Although Borlaug was modest about his award, by that time he had adopted a "missionary zeal" for increasing world food production and decreasing global population.[64] Scientists from the Middle East whom Borlaug trained became known as Borlaug's "wheat apostles."[65] And Borlaug's colleagues recalled him preaching, "What Mexico did, your country can also do, except that yours should do it in half the time."[66] Borlaug, though trained as a plant pathologist, gained a new status as one of the most respected wheat breeders in the world and used that platform to spread his gospel. Borlaug was not shy about making the link between widely adapted varieties and global food production. In an undated outline of a report titled "The Development of High Yielding, Broadly-Adapted Spring Wheat Varieties," Borlaug handwrote the rest of the title to be "and its Significance for Increasing World Food Production."[67] In the margins of the outline, he wrote "KF" and "CK" next to various sections. These were Keith Finlay and Charles Krull, Borlaug's two colleagues who were critical to promoting wide adaptation as a plant breeding ideal.

CHARLES F. KRULL AND THE ROCKEFELLER FOUNDATION'S COOPERATIVE PROGRAM IN THE MIDDLE EAST, 1965–1968

Charles F. Krull was a cereal breeder for the RF in Colombia from 1960 to 1965 and in Mexico from 1965 to 1968. Krull was a crucial advocate of Borlaug's concept of wide adaptation, especially with scientists in the Middle East. Krull also led the analysis of the first few International

Spring Wheat Yield Nurseries. While Borlaug was busy traveling, Krull served as Borlaug's program manager, editor, and proxy in Mexico. The records created by Krull in the late 1960s, including his correspondence, trip diaries, and an oral history, provide a unique insight to the RF's program, goals, and personalities.

Krull applied to work with the RF directly out of graduate school at Iowa State University, where he had worked with Kenneth J. Frey, a well-known oat breeder. The RF was looking for a cereal breeder to work in their Colombian Agricultural Sciences program, and Krull fit their requirements. Arriving in Bogotá, Colombia, in June 1960, Krull worked with the RF's wheat breeder John Gibler.[68] Krull and Gibler both became involved mainly in the wheat improvement program in Colombia, with oats and barley as secondary areas of focus.[69] After a few years, however, the RF considered phasing out the Colombia program due to successful training of several Colombian scientists.

In the mid-1960s Borlaug needed assistance with the Mexican wheat program as he took on a more international role. Borlaug also needed help analyzing results of the international wheat yield trials. For several years, only preliminary results had been sent to the international collaborators.[70] Borlaug needed someone with experience in both plant breeding and statistics to help him, and Krull was experienced in both from his dissertation work. In August 1965 Krull transferred to Mexico to coordinate the international wheat yield nursery and its analysis, as well as to cover many of Borlaug's duties in Mexico while Borlaug traveled. Krull was named resident coordinator of International Wheat Program in May 1967. Gibler, meanwhile, was transferred to Ecuador to continue working on wheat there.

Having Krull in Mexico was a boon to Borlaug's program on wide adaptation. With the analyzed results of the International Spring Wheat Yield Nursery, Borlaug now had empirical evidence to support wide adaptation: several of the Mexican varieties yielded, on average, the best of all varieties tested. Borlaug stated in his 1967 oral history: "We begin to understand some of the basic things that underlie this adaptation. This, to me, is a fundamental discovery that has long been overlooked. And it has been borne out now, and we have ample evidence, some of which has been reported in these recent bulletins that Dr. Krull has been getting out, that are backed up by large quantities of experimental data."[71] Krull's analysis of the international wheat nursery results bolstered Borlaug's confidence to expand the RF's wheat program into the eastern hemisphere.

Throughout his time with the RF in Mexico, Krull consistently stated that scientists should consider the importance of widely adapted wheat varieties, that countries should focus efforts on only one breeding and testing program for fertilized and irrigated environments, and that widely adapted varieties chosen under favorable environments could unequivocally outperform local varieties, regardless of environment. These views were not mainstream among wheat scientists, especially those from the FAO who were working in the Middle East.

Krull often argued that wide adaptation was an important and undervalued concept in wheat breeding. Speaking on the "elusive concept of breeding for adaptation," Krull addressed the Minnesota-based Crop Quality Council in 1967 about the "deeply ingrained philosophy that is held and taught by most of the North American graduate schools that such adaptation is probably neither possible nor desirable."[72] Krull had written earlier: "Plant breeders frequently feel that varieties must be well adapted to only very small areas. They feel that since variety × location interactions are frequently encountered the ideal variety must be narrowly adapted. Indeed, such varieties can be produced. *It is also possible, however, as is illustrated by these data, to produce varieties that are widely adapted.*"[73] Krull obviously disagreed with mainstream plant breeders that varieties should be bred for local conditions. He even pondered the "possibility of producing spring wheat varieties with nearly universal adaptations."[74] Krull certainly did not lack Borlaug's missionary zeal.

Although Krull traveled to the Middle East only a few times, he frequently wrote to two FAO scientists working in the Middle East: Abdul Hafiz, a regional consultant for the FAO's Near East Wheat and Barley Improvement Project who was located in Egypt in the 1960s and 1970s, and C. L. Pan, a cereal breeder for the FAO in Iraq who, like Borlaug, had studied at the University of Minnesota.[75] Hafiz also helped coordinate the Near East–American Spring Wheat Yield Nursery with Borlaug.[76] The RF was interested in working in the Middle East and continuing their collaboration with the FAO, but scientists from the two organizations had different crop-breeding philosophies. The FAO team held the traditional position that crops needed specific adaptation to local conditions. Krull, on the other hand, attempted to influence wheat breeders in the Middle East to adopt breeding and testing practices more like Borlaug's methods.

Unlike the irrigated Sonora region of Mexico, where farmers clamored for semidwarf wheat, the Middle East had a diversity of wheat-farming practices. In the 1960s, plant breeders in the Middle East focused on low-fertility conditions that farmers were most likely to experience. Krull, like Borlaug, argued that wheat breeding should focus on only highly fertilized conditions. Krull made a trip to the Middle East in April and May 1966, where he recorded his detailed observations and opinions of the wheat programs there. Krull observed that in a dryland area of Jordan, "the yield nurseries showed a decided lack of fertilizer, and this tended to make all varieties look the same. The reasoning was that most of the farmers do not use fertilizers so varieties must be selected under these conditions."[77] He felt, however, that "this is a common fallacy among wheat breeders in under-developed countries, and there is actually little basis for it."[78] Krull reasoned that well-fertilized environments allow the breeder to see the variability between varieties to help them make their selections. In a letter to Hafiz in 1966, Krull wrote, "As suggested, I would like to see the nurseries more heavily fertilized. It is simply much easier to see yield differences at these high fertility levels. Putting on a good amount of fertilizer tends to iron out any soil differences that there might be, so that the differences in yields observed are mainly genetic."[79] Krull was consistent and persistent in his argument for high fertility and testing.

While visiting Iraq on the same trip, Krull wrote, "The experiments needed fertilizers badly and there were water logged spots that damaged parts of most experiments. . . . Pan had not fertilized the nursery on the basis that farmers do not fertilize."[80] After some discussion with Pan, Krull thought that he "finally seemed pretty well convinced" to use higher levels of fertilizer.[81] Pan indeed seemed convinced. He reported on the visit to his former advisor, the esteemed plant breeder Herbert K. Hayes. His experiments were conducted under the "local method of farm management with a brief that any promising varieties thus screened out will be adoptable to the local conditions. Dr. Krull's way of thinking in this respect, however, is quite different from mine. He thought that such a variety trial should be carried out in a field provided with the best conditions for the growth of the plant."[82] He continued, "This seems to me a more realistic way of approach, and I am prepared to follow such new approach when I design trials in the future. . . . I would become much more convinced if you also can endorse this new approach."[83] Unfortunately, Hayes's response is not included in the archives.

Krull also argued that scientists should breed and select plants under high fertility. He wrote to the FAO's Hafiz that "if the breeder is only working at the fertilizer level now used by the farmers, by the time the variety is actually selected and multiplied, it will already be obsolete with the better farmers."[84] Krull and Borlaug both felt that wheat breeders should anticipate higher fertilizer levels in the future and breed for responsive varieties. Hafiz echoed this, writing to Krull, "No doubt, the Cereal Breeders have now realized the great importance of breeding and testing varieties under high fertilization . . . the Breeders will have to cater for varieties suitable to be grown under high fertilization, which is the only answer to meet the food shortage."[85] Thus, it appears Krull influenced the thinking of both Hafiz and Pan around fertilizers.

Krull reflected on a trip to the Middle East in 1966 that, "there seems to be little basis for the widely spread belief that varieties selected under high fertility do not usually do as well under low fertility."[86] Krull drew from the results of the international nurseries to argue against this belief. In the *Results of the Fourth Inter-American Spring Wheat Yield Nursery*, published in 1967, Krull and his coauthors challenged the prevailing idea that "each environmental niche must ideally have its own set of varieties" with the finding that the Mexican varieties had the highest average yields around the world.[87] In his presentation to the Crop Quality Council, Krull argued:

> If we seed 10 Mexican and 10 Indian varieties without fertilizer in India, we find that they all yield about the same. If we then seed the same experiment at another site with 120 pounds per acre of nitrogen, we find that the group of Mexican varieties yields considerably more than the tall, weak-strawed Indian lines. . . . The varieties that yield well with fertilizer also tend to be the same ones that yield best with poor management. This is very nicely illustrated by . . . literally hundreds of smaller tests that were run last year throughout India and Pakistan, and to a lesser extent in other countries in the Near East and the Americas.[88]

He stated further, "My point is that the presence of variety × location interactions does not necessarily imply that the same varieties are not the highest yielding in all environments."[89] In other words, Krull argued that wheat could be widely adapted across not just locations but also diverse environmental conditions.

Krull also extended his argument to soil moisture, arguing that one variety could also be the best performer in both irrigated and rainfed environments. He said to the Crop Quality Council, "Evidence is accumulating that this same thing is true in irrigated versus dryland conditions. . . . Such a statement is considered to be rank heresy by most wheat breeders."[90] Finally, he argued "that varieties that show good adaptation in area are also better adapted over time," meaning they had consistently high yields year after year.[91] Krull wrote in 1965 that "the published results of our first five international yield trials have shown that it is possible to produce a series of varieties that are capable of outyielding local varieties from Chile to Canada and from Minnesota to the Near East."[92] He believed, like Borlaug, that high yield and wide adaptation made the Mexican semidwarf wheats superior to nearly all other wheats, no matter their environment.

Krull's thoughts on breeding for soil moisture echoed his opinions on soil fertility. Krull argued that varieties selected under irrigation could still be adapted to moisture-stressed environments, and that they were superior to local varieties. He wrote to Hafiz in 1966, "It appears that varieties that are adapted to intensive irrigation may also be adapted to very droughty conditions. Thus, it is not necessary to initiate a separate program for the irrigated and arid areas."[93] In a 1967 letter to Byrd C. Curtis, a plant breeder at Colorado State University, Krull wrote that the Mexican semidwarf wheats were "extremely productive under irrigation and high fertilization, but the results of our international nurseries indicate that they do as well as supposedly drought-resistant varieties under poor conditions."[94] He wrote further that "in other words, the dwarfs *respond* to but do not necessarily *require* irrigation and extremely heavy fertilization."[95] This argument implies that widely adapted varieties have an inherent (or genetic) high yield, that they can efficiently use moisture and nutrients under both surplus and scarcity.

While Hafiz and Pan were both amenable to Krull's fertilizer suggestions, they disagreed with his recommendations for dryland agriculture. Hafiz wrote to Krull that agronomic improvements ("agrotechniques") were necessary for dryland conditions, not just widely adapted varieties: "For dry farming areas we will try to follow your suggestions but still I feel these areas require at least one comprehensive programme for the Region not only from the point of view of developing drought resistant and higher yielding varieties but also for developing better agrotechniques for the efficient use of soil moisture and fertilizers. . . . It is really a very big and very difficult problem, but at the same time the most important and immediate one."[96] Krull responded: "I certainly do not disagree that it would be worthwhile to concentrate heavily in at least one place on drought resistance. My point was simply that I don't believe it would be wise to separate it from an irrigated program as it appears to be possible to produce drought resistance varieties that are also adapted to irrigated conditions."[97]

Pan also wrote to Krull about the problems of dryland farming. For the wheat-growing areas of Iraq, Pan wrote, "It seems that wheat breeding should concentrate on drought resistance in the north and salinity tolerance in the south."[98] A year later, Pan still insisted to Krull that a drought-resistance was critical in Iraq. He wrote, "As you know more than two thirds of the wheat crop in Iraq are grown in the north in the rainfed area. But rainfall varies very greatly from year to year. It seems that the most effective way to increase the yield level of wheat in the rainfed area is to use drought resistant variety."[99] Here, Pan touched on a decades-long scientific debate on the efficiency of selection environments, which will come up in the next few chapters.

During the mid-1960s the RF's wheat program decided to focus on irrigated areas because they could raise yields more easily there. This strategy was a clear contrast with the FAO program in the Middle East. The FAO breeders evidently held a different philosophy of agricultural development from the RF wheat scientists. While the FAO focused on improving agricultural production under all conditions, the RF was "betting on the strong" and emphasizing production gains in irrigated and fertilized areas. Krull wrote, "While there is interest in many countries in producing varieties that do not require fertilizer or water, *there is no such group of varieties.* The important thing in changing the production pattern in a country is to introduce varieties that will *respond* to good management and then change the management."[100] This statement reflects a belief, held by the RF administration and Borlaug, that technical change would inevitably lead to social change. Borlaug and Krull viewed "good" agronomy as maximizing yield under high-resource conditions, while others might define it as getting by with the resources at hand.

Krull left Mexico in 1968 owing to a divorce, but remained affiliated with the RF.[101] While Krull seems to have been very influenced by Borlaug, Krull left an impression on Borlaug as well. Borlaug used Krull's data analyses to support the spread of widely adapted, fertilizer-responsive wheats. Krull argued that the most productive way to improve a national plant-breeding program was to aim for widely adapted varieties selected under favorable environments. His evidence was the results of the International Spring Wheat Yield Nurseries. Around the same time, Keith Finlay used empirical analysis to take Borlaug and Krull's results a step further: to quantify adaptation across environments.

KEITH FINLAY'S CORRESPONDENCE ON ADAPTATION, 1963–1968

Agriculturalists had long regarded adaptation as a factor that could not be predicted or quantified, but only tested through trial and error by introducing plant varieties to new locations. Starting in the late 1930s, scientists began using analysis of variance models to analyze crop performance against independent variables.[102] These models could, for example, show that a variety's phenotype changed based on location or experimental treatment. Then in 1963 an Australian wheat breeder, Keith W. Finlay, and his colleague, statistician Graham N. Wilkinson, created an experimental design and mathematical model that measured the phenotypic stability—or adaptation—of plant varieties in different environments.[103] The model was a simple logarithmic plot of a variety's yield versus the mean yield at a location; in other words, performance versus an environmental index. The model became immediately popular among plant breeders and led to a variety of other "stability models" that are still employed today. As one of the first computational analyses of plant breeding, this model influenced plant breeders to study crop adaptation to environments.[104] The prominent crop physiologist Lloyd T. Evans called stability models "the plant breeder's icons, ubiquitous but with a variety of styles to support a variety of dogmas."[105]

Finlay was a professor of plant breeding at the Waite Agricultural Research Institute at the University of Adelaide, Australia. He provided an academic counterbalance to Borlaug, though he shared many of Borlaug's goals. Borlaug became aware of Finlay through Vogel, who considered Finlay a "first choice" hire to coordinate the RF's Indian wheat program.[106] Finlay visited Borlaug from October through November 1963, partly for an academic exchange and partly to express interest in an open wheat breeder position in Mexico. He presented his work on adaptation while touring Mexico and Colombia. Despite finding Finlay "a very capable theoretical research scientist," Borlaug found him too academically oriented for either the India or the Mexico position, where Borlaug wanted someone with an inclination toward fieldwork.[107]

A few months later Finlay wrote to Borlaug to apprise him that he had submitted a research proposal to study adaptation together with the RF's scientists in Mexico. Robert Osler, the assistant director of agricultural sciences for the RF, let Finlay know that the success of the proposal depended largely on how Borlaug prioritized it. The RF rejected the proposal in September 1964. As soon as 1965, however, Borlaug proposed bringing Finlay back to Mexico to help Krull set up to analyze the international trial results. It appears that Finlay was able to visit in 1966 and in 1967, and Borlaug or Krull provided him with data from the international trials to analyze for his own research.

Borlaug wrote to Finlay in 1964, "Since I last saw you we have learned considerably more about adaptation of the Mexican breeding material in far-away places. . . . The Mexican material was equally as well adapted in India as in Sonora."[108] Finlay responded, "There is certainly no doubt that the more recent Mexican varieties have a very wide adaptation," and he hoped they could continue working on adaptation together.[109] Finlay also included some preliminary analyses of the 1961–1962 and 1962–1963 Near East–American Spring Wheat Yield Nurseries, where he plotted the varieties' average stability by their average yield, clustering the varieties into groups. He found that the newer Mexican varieties were superior in terms of stability across locations and having a higher average yield, although there was not much difference between the varieties released in 1960, 1962, and 1964.[110] In other words, there was not much different between the tall and semidwarf varieties: both were widely adapted and high-yielding.

Borlaug wrote to Osler the same day he wrote back to Finlay. He wrote, "I feel that Dr. Finlay has developed some useful information to partially explain adaptation phenomena we have already uncovered in the FAO-Near East-American Spring Wheat Yield Tests, and the Inter-American Spring Wheat Yield Nurseries."[111] Meanwhile, Louis P. Reitz, who led the USDA's wheat research, also corresponded with Osler about Finlay. Reitz wrote that Finlay's analysis "surely would lead to wider use of the fine Mexican materials and the work might lead to improved pools and greater understanding of gene pools. Some benefits would come even if the work merely 'proved the obvious.'"[112] Finlay's analytical work appeared useful to "prove" the wide adaptation of Borlaug's wheats.

In late 1966 Finlay wrote a long, detailed letter to Borlaug about adaptation, the analysis of the international yield trial results, and the future directions of CIMMYT. Though excited about CIMMYT's expanded international programs, Finlay also had some reservations about Borlaug's research program. He wrote to Borlaug, "Although this wide adaptation is one of the strong points of your programme, it is also possibly the weakest!"[113] He suggested that Borlaug should collect more basic data to determine what causes wide adaptation, writing: "Your present wide adaptation is resulting from selection successively in a number of different environments, but the *type* and *degree* of adaptation is not known for any particular variety until it goes into the International Yield Trial."[114] Finlay thought that more testing throughout the breeding process would be helpful. He cautioned Borlaug to understand more about the mechanism of wide adaptation before advancing too quickly with his international wheat program.

Finlay also had some concerns about Borlaug's shuttle breeding method, writing, "The selection technique used at present certainly allows the selection of widely adapted genotypes but it also *automatically eliminates* genotypes with exceptional potential for yield given the correct specific environment."[115] Thus Borlaug's program might be weeding out varieties well adapted to conditions such as drought. Finlay suggested separating the breeding of rainfed and irrigated wheat varieties to "exploit both sets of environments much more efficiently by having varieties which are widely adapted to environments *within* each set."[116] Yet Finlay believed that with some experimental modifications, Borlaug's work could "revolutionise thinking in plant breeding circles."[117]

Finlay was meanwhile working with Australia's well-known plant breeder Otto H. Frankel to promote the conservation of plant biodiversity. They worked together on the International Biological Program project called Biology of Adaptation, which Finlay convened starting around 1966. The International Biological Program (IBP, 1964–1976) was an attempt at "big biology" to collect large-scale data sets, modeled after the International Geophysical Year.[118] The Biology of Adaptation project and another on "plant-germ-plasm pools," chaired by Frankel, fell within the IBP's subcommittee on "Use and Management of Biological Resources."[119] Finlay and Frankel were not unusual in their interest in plant biodiversity; plant biodiversity conservation became a major focus of plant breeders around the world, including India's famous M. S. Swaminathan, who was also involved in the IBP program on adaptation.

The original goal of the Biology of Adaptation project was an "analysis of the performance of a large number of varieties in certain standard, selected environments . . . and consequent analysis of productivity in genetic, physiological, and ecological terms" for four to six crops in an experiment like Borlaug's yield trials.[120] Although Borlaug and Finlay appeared to have a cordial relationship, Borlaug was initially unimpressed by the IBP's Biology of Adaptation project. On his copy of the "IBP Second Circular" from August 1966, Borlaug wrote in the margins of the planned experiments, "Being done by RF," "Charlie—this looks like our own ISWYN [International Spring Wheat Yield Nursery]," and, regarding Finlay as coordinator for temperate zone cereals, Borlaug wrote, "Competition?"[121] Borlaug wrote to the RF's director of agricultural sciences, Sterling Wortman, "Why should we set another organization up in competition with our own?"[122]

In fact, the resemblance of the projects was likely due to Frankel himself, who favored Finlay's analytic aspects of adaptation along with Borlaug's practical aspects.[123] But Borlaug, ever focused on expanding his wheat program, was offended rather than flattered. Frankel wrote to Borlaug, "We are mainly concerned with a broad adaptability study on the Finlay pattern; you are, I imagine, mainly concerned with the agricultural success."[124] Frankel became personally interested in recruiting Borlaug to the IBP adaptation program and invited him to the conference meetings in Rome. By January 1967 Borlaug appeared to be on board to support the IBP's adaptation program. CIMMYT collaborated with IBP to conduct adaptation experiments as part of CIMMYT's Sixth International Spring Wheat Yield Nursery of 1969–1970.[125] The IBP wheat adaptation program did not seem to progress much beyond that, however, and likely was simply subsumed by CIMMYT's existing international yield nurseries when Finlay started working there in late 1968.

Despite his earlier rejections by the RF and Borlaug, Finlay helped bring Borlaug's program on adaptation to international academy. He brought Borlaug to the Third International Wheat Genetics Symposium, held in Canberra, Australia, in early August 1968, to give a keynote titled "Wheat Breeding and Its Impact on World Food Supply."[126] This conference signaled Borlaug's wider acceptance by the wheat research community. Finlay also presented a paper titled "The Significance of Adaptation in Wheat Breeding."[127] He used the results of Borlaug's international trials to show that varieties could be bred with both high average yield and wide adaptation.

Though Borlaug had passed over Finlay for positions at CIMMYT several times already, after Krull's departure in 1968 Borlaug needed someone with a strong mathematical background to help with the international trials and general administration of the wheat program.[128] Gibler was promoted to associate director of the wheat program, and Finlay was recruited to assist him and Borlaug. Finlay was quickly hired as "Director, Basic Research and Training (International nurseries and data retrieval)" for the maize and wheat programs at CIMMYT and remained there until his death in 1980.[129]

Finlay's work on adaptation, both theoretically and programmatically, helped solidify it as a measurable object of study in the plant breeding community. Corresponding with Borlaug starting in 1963 and working at CIMMYT for a dozen years, Finlay "proved the obvious" of Borlaug's adaptation program—that certain varieties could be widely adapted across environments—through his analysis of adaptation.[130] The two scientists were not completely in sync in their views on wide adaptation, however. Finlay called for more understanding of the mechanisms of adaptation, while Borlaug focused on rapidly growing his wheat program. Finlay appeared more interested in how adaptation emerged and how it could be developed in a plant-breeding program, especially drawing on plant diversity. Borlaug, on the other hand, seemed more concerned with the practical and immediate uses of widely adapted varieties, and ignored empirical evidence at times. Despite their differences, Borlaug and Finlay depended on each other for theoretical models and experimental data, which they both used to promote wide adaptation internationally.

ADAPTATION WITHOUT CONTEXT

Borlaug, Krull, and Finlay were three influential figures in international wheat research in the 1960s. Borlaug undisputedly played the major role in elevating wide adaptation as a goal in agricultural science and establishing the narrative and meaning of wide adaptation. Krull and Finlay, however, have been rather overlooked in the history of agricultural science. Krull promoted wide adaptation and breeding for ideal environments in the Middle East. Finlay, on the other hand, corresponded with Borlaug about the more theoretical aspects of wide adaptation. He also promoted Borlaug's wide adaptation through international research forums. Finlay's theoretical and administrative work on adaptation helped solidify it as a measurable object of study in the plant sciences. Best known for his mathematical model of adaptation, Finlay started a revolution in quantitative plant breeding.

It is unquestionable that the Mexican semidwarf varieties combined several genetic qualities that allowed very rapid international adoption. The daylight insensitivity, dwarfing genes, and rust resistance of these varieties added to the intrinsic wide adaptation of wheat. These varieties had, on average, high yields regardless of the location and agronomic conditions. They also fit the RF's mission of helping countries that lacked a well-developed agricultural research system.

However, the international consequences of Borlaug's program are not all positive. Although Borlaug's wheat varieties garnered international interest, critics began pointing out that these varieties did not always perform well in rainfed or low-fertility environments. Borlaug more or less ignored these claims and focused on the dire consequences of traditional agriculture and overpopulation. He relied heavily on the averaged, decontextualized results of his international trials and brushed aside anecdotal evidence from field staff in the Middle East. The more he was criticized, the more he sank into his position that widely adapted wheats could outyield local varieties even under rainfed or low-fertility environments. Simultaneously, Borlaug's research program worked exclusively under high-fertility conditions to maximize the varieties' response to nitrogen fertilizer. Thus, while wide adaptation itself is not inherently problematic, Borlaug tied wide adaptation to the need for high fertility in a way that ignored the reality of many farmers around the world who lacked access to fertilizers.

Borlaug and Krull promoted wide adaptation to expand the RF's wheat programs and to increase global wheat production, but they did so under questionable scientific premises. The team promoted breeding and testing under only high-fertility and irrigated conditions but extended the meaning of wide adaptation from adaptation across location to adaptation across agronomic conditions. While the results of the international yield trials on average supported their assumptions, Borlaug did very little, if any, investigation into the performance of his wheat varieties under farmers' conditions outside of Mexico. This is important because yield trials are often biased as a consequence of more careful management, better soil conditions, and so on, and results from bad years (such as drought) are often thrown out. Borlaug and Krull also drew firm conclusions about the superiority of the semidwarf wheats while comparing these to only a few local varieties, including varieties that were photoperiod sensitive and thus would not yield well outside of their growing zone. Borlaug and Krull did not examine the possible biases of the international yield trials, even when confronted with alternative explanations.

Finlay's involvement with Borlaug and CIMMYT points to some problems with Borlaug and Krull's mission-driven approach to expanding the RF's wheat program. Namely, Borlaug and Krull focused on irrigated and fertilized conditions through controlled experiments while overlooking the genetic and physiological factors that contributed to wide adaptation. Borlaug downplayed the genetics of photoperiod insensitivity to emphasize how his method of shuttle breeding led to widely adapted varieties. This breeding technique is still employed by CIMMYT to select for wide adaptation.

We can say that Borlaug made a series of reductionist arguments for wide adaptation. Modern scholars have examined the phenomenon of "disembedded grain," and this descriptor seems apt.[131] Borlaug developed his international program on wide adaptation without much engagement with farmers outside of Mexico; even in Mexico, he worked with wealthier farmers who used irrigation and fertilizers. While there is nothing inherently problematic about introducing plant varieties to new locations, there is a problem when these varieties require a set of agronomic techniques vastly different from the local context.

CHAPTER 2

PROPER AGRONOMY

THE INDIAN CONTEXT OF A NEW PLANT BREEDING IDEAL, 1960–1970

Wheat cultivation in India changed radically in the 1960s due to new technologies and policy reforms introduced during the Green Revolution. Scholars have thoroughly studied the technological and political changes of this period.[1] Absent from most discourse on the Green Revolution in India, however, is that in the mid-1960s Indian scientists adopted a new plant-breeding philosophy—that crop varieties should be widely adapted and that research should be conducted under ideal agronomic conditions. Many people have criticized the Green Revolution for its unequal spread of benefits, but few of these critiques address wide adaptation. I argue that Indian agricultural scientists use wide adaptation to implicitly justify their focus on highly productive land while ignoring marginal or rainfed agriculture. So long as wide adaptation is celebrated, scientists and planners can brush aside concerns about social equity and justice.

Norman Borlaug was the major force behind this change in plant-breeding philosophy in India. Borlaug and the Rockefeller Foundation (RF) became reformers of the Indian wheat research system in the 1960s. The Indian wheat program, under RF influence, underwent three significant changes during this time. The wheat program became centralized, scientists increased the fertilizer levels in field trials (using fertilizer rates multiple times higher than average field in India), and scientists judged varieties on their average performance rather than their fit to a specific location or condition. The RF also introduced the Mexican wheats to India on the premise that existing Indian varieties would not increase national wheat production. They claimed, ostensibly to placate India's economic planners who favored a more socialist agricultural system, that widely adapted varieties would still produce high yields in marginal environments. In reality, the Indian wheat program became focused on the favorable agroclimatic conditions of northwest India.

In this and the next two chapters I examine the consequences of the wide adaptation regime in India and other countries. I argue that administrative changes in the mid-1960s have created a research system that is "locked in" to wide adaptation. Not all Indian scientists agree with the philosophy of wide adaptation, but they are constrained by the system's commitment to it. A wheat research and testing system based on wide adaptation severely limits its ability to help small and marginal farmers, to release better-adapted varieties, and to address climate change adaptation. Understanding the history of wide adaptation in India lends some insight to why it became so entrenched in the current system.

A BRIEF HISTORY OF INDIAN WHEAT RESEARCH

The history of agricultural research and production in India is closely tied to British colonialism. Famines frequently occurred in British India due to a combination of climatic fluctuations and ineffective famine prevention or relief.[2] The British Raj focused on improving cash and export crops over food crops, which may have contributed to the decline in crop yields starting around 1920.[3] In 1942, with the start of World War II, the British government announced a "Grow More Food" campaign in India, encouraging farmers to switch from cash to food crops.[4] Grow More Food focused on increasing cereal grain production and availability of irrigation.[5] Ultimately, the campaign languished as a result of lack of investment and concrete plans. The Bengal famine of 1943 is a particularly bleak outcome of colonialism in which up to three million people starved to death as Winston Churchill diverted food away from the region.

The British established the first formal agricultural research institute in India, the Imperial Agricultural Research Institute (IARI), in 1905 in Pusa, Bihar.[6] In 1908 five agricultural colleges were established. By 1910 India had special government laboratories for wheat and rice.[7] The administrative body of the Imperial Council of Agricultural Research (ICAR) was formed in 1929. IARI moved to Delhi in 1934 after an earthquake at Pusa, and Delhi became the central seat of Indian agricultural research.[8] Punjab was also a major research site, but mostly in the western region that would become Pakistan after the partition. After India's independence in 1947, the Imperial Agricultural Research Institute and

Imperial Council of Agricultural Research changed their names to the Indian Agricultural Research Institute and Indian Council of Agricultural Research, respectively, but retained their abbreviations.

Before the Green Revolution, wheat cultivation in India was mostly limited to the northwest, while rice prevailed in the east and south, and millets in the peninsular region.[9] Indian wheat-breeding programs existed from the early 1900s, but efforts were decentralized and resulted in marginal gains in wheat yield. Contributing to the marginal gains was the fact that foreign wheat varieties were not well adapted to Indian conditions.[10] This was likely attributable to India's warmer climate requiring spring wheats. Sir Albert Howard and Gabrielle L. C. Howard, who worked at the Pusa station, are known as the pioneers of Indian wheat research. They wrote in their 1909 book that "the introduction of exotic wheats into India has been a long record of failure."[11] The Howards also noted the location specificity of agricultural research, writing, "The smallest differences in procedure are closely bound up with differences in local conditions."[12] Despite early failures, India systematically experimented with foreign plant introduction from 1942 onward.[13]

Both before and after independence, Indian wheat scientists lobbied for ICAR to support their research. Indian wheats face, then and now, three different types of rust, a fungal pathogen that attacks wheat and can reduce its yields: leaf rust, stem rust, and stripe rust. Indian scientist Benjamin Peary Pal coordinated wheat disease research in India throughout the 1930s, 1940s, and 1950s.[14] Pal received his PhD in botany from the University of Cambridge in England under Rowland Biffen and Frank Engledow. He also mentored India's most celebrated wheat scientist, Mankombu Sambasivan Swaminathan. In 1934 Indian scientists decided "that a collaborative beginning for breeding rust-resistant varieties of hill wheats should be undertaken at Simla . . . placed under the charge of Dr. B.P. Pal."[15] ICAR approved Pal's coordinated scheme to control rust in 1952, two years after he was promoted to director of IARI.[16] A coordinated agricultural program would soon become a central feature of Indian agricultural research, as Pal became director of ICAR in 1965.

By 1957 introducing and adapting foreign wheat varieties to Indian conditions was still mostly unsuccessful, which reinforced the idea that research should be locally targeted. Indian agricultural scientists S. M. Sikka and K. B. L. Jain wrote in 1960 that yield is "best when the plants of a particular variety are in harmony with the environment. It is now well recognised that a single variety of any crop, howsoever improved it may be, cannot be a universal success."[17] Sikka was head of IARI's botany department at that time, so we can assume that his views were mainstream. Sugarcane breeder T. S. Venkataraman showed skepticism toward the concept of wide adaptation in a 1953 ICAR advisory board meeting. But Pal rebuked Venkataraman, showing his support for wide adaptation before having close interactions with Borlaug. He said that for crops like sugarcane, "it was true that the crosses were best made in certain localities because these crops did not produce flowers in all localities. . . . But in the case of crops like wheat and rice their experience was that that did not happen."[18] Pal argued that the physiological aspects of wheat made it particularly adaptable compared to other crops, and that a variety developed in one state could benefit the entire nation. Thirteen years later, Pal was validated as India imported eighteen thousand tons of Mexican wheat seeds to be planted across India.

AGRICULTURAL POLITICS AFTER INDIA'S INDEPENDENCE

India became independent from Britain in 1947, and Prime Minister Jawaharlal Nehru was the country's major political force until his death in 1964. Under Nehru's leadership, India's policies veered toward a centralized and socialist agenda. India's constitution prescribed that "the ownership and control of the material resources of the community are so distributed as best to subserve the common good" and should not result in the "concentration of wealth."[19] Nehru was also a strong proponent of modernism and "the scientific temper."[20] Nehru's modernist view of science and technology would have profound impacts on agricultural policy.

India's independence brought new ambitions for self-reliance in food and agriculture. Economic planners hoped to separate India's agriculture from its colonial past. Historian Benjamin Robert Siegel described how the Bengal famine profoundly influenced national policies around food production and distribution.[21] Agricultural reforms included socialist land reform, helping landless peasants settle new land, and abolishing absentee landlords.[22] Yet India's first five-year plan, in 1951, focused more on industrial production than on agriculture. This is because Nehru and his advisors believed that industrial growth would stimulate the economy, provide jobs, and prove the country as modern.[23] Nehru also thought that focusing on industry would create more demand for agriculture and therefore boost production.[24] The first five-year plan also focused on village self-sufficiency and community development—the development model that would prevail until the RF's involvement in the mid-1960s.[25]

The second five-year plan (1956) expanded support for agriculture and social welfare programs, but it reduced national spending on fertilizers. India imported its chemical fertilizers, which required spending valuable foreign exchange.[26] Thus, the second plan aimed to use the scarce foreign exchange supply for industrial purposes and hoped that domestic fertilizer production would increase.[27] Then, as the result of a crop failure, in 1957 the Indian government set up the Food Grains Enquiry Committee.[28] This committee recommended a shift to an agricultural development approach focused on high-production areas, including application of chemical fertilizers. India's Planning Commission rejected those recommendations because they went against Nehru and the Planning Commission's main goal of social equity.[29] In the third-year plan (1961), the final plan before the Green Revolution, the Planning Commission put a high priority on self-sufficiency in food grains, but still reduced spending on irrigation and fertilizer.[30] In the time between independence and the Green Revolution, Indian planners struggled over how to improve domestic production and consumption of food crops. While rationing and food-related campaigns were common, it is worth noting that India did not experience famine during this period.

FOUNDATION INVOLVEMENT IN INDIAN AGRICULTURAL RESEARCH

US foundations became involved in Indian agricultural development in the 1950s, after India's Grow More Food campaign ended. The Ford Foundation, the RF, and the United States Technical Cooperative Administration (which became the US Agency for International Development in 1961) all involved themselves in agricultural development in India in the 1950s and 1960s. Initial programs focused on community development through village-centric programs and technical assistance through demonstrating new practices and technologies.[31] Postindependence India also consumed surplus grains that the United States distributed to India through Public Law 480 (PL-480) starting in 1954.

The RF and other foundations viewed the Punjab region, in northwest India, as especially fertile ground for both social and agricultural development experiments.[32] The Rockefeller Foundation and the Harvard School of Public Health conducted a large-scale study of population control in Punjab rural villages from 1953 to 1969, known as the Khanna Study.[33] The researchers remarked, "India . . . is the cauldron in which mankind will be tested."[34] Many US scientists viewed India as a laboratory for democracy and the self-help ideology that prevailed during the Johnson administration.[35]

The RF got involved in Indian agriculture in 1952, when they sent three scientists to survey farming in India. Then in 1954 the Indian government contracted two RF maize scientists, Edwin J. Wellhausen and Ulysses J. Grant, to survey India's maize system and to advise on collaboration between India and the RF. At that time, Wellhausen directed the Mexican Agricultural Program (MAP) and Grant led the RF's Colombian maize improvement program. The RF and government of India signed a memorandum of understanding in 1956 aimed to improve secondary education in agriculture and focused on three cereal crops: hybrid maize, sorghum, and millet. The Indian government worked with the RF on two Joint Indo-American Councils in 1954 and 1959 to develop agricultural education and research policy for India.[36] The first Joint Indo-American Council recommended that India implement a rural agricultural education system based on the US land grant colleges.[37] The RF supported this project in partnership with five US land grant colleges.[38] On March 8, 1957, Ralph W. Cummings, a soil scientist from North Carolina State University, arrived in India as field director for the RF and Grant started as the assistant field director and director of maize breeding.[39] With much direction from the University of Illinois, the RF helped establish an agricultural university in Pantnagar, Uttar Pradesh, in 1960 and six more universities were set up in the following five years.[40]

India was also involved in the US Technical Cooperation Program starting in 1952. This program distributed and demonstrated fertilizers and soil fertility testing to Indian farmers. Frank Parker, an American agronomist stationed in India in the early 1950s with the Technical Cooperation Program, stated, "India is not overpopulated, it is underfertilized."[41] A major reason for the United States' involvement in India's food supply was to promote social stability in the region. US foreign policy experts believed that food and agricultural aid were a tool of national security, made explicit in the title of President John F. Kennedy's "Food for Peace" law in 1961. As in Mexico, the RF worked as a development agent to promote economic and political stability.

The RF built on previous efforts by the Ford Foundation to increase food production in India. In 1961 the Ford Foundation had started the Intensive Agricultural District Programme, also known as the Package Programme, which aimed to improve rural development in select districts through fertilizer, irrigation, and modern crop varieties. The Ford Foundation specifically worked in areas with assured irrigation or adequate rainfall. American experts involved in the program found that Indian crop varieties were not responding to improved agronomy and "the major criteria for selection of a variety for multiplication should be its ability to give higher yields under cultivator's conditions including responsiveness to heavy doses of fertilizer and ample water and plant protection."[42] The RF would pick up where the Package Programme ended.

With RF support from the 1960s onward, most wheat research efforts in India focused on the northwest region of Punjab, Haryana, and western Uttar Pradesh, which make up an agriculturally productive and mostly irrigated region. Most of the prominent research organizations were in the northwest, including IARI, in New Delhi; G. B. Pant University of Agriculture and Technology, in Pantnagar; and Punjab Agricultural University, in Ludhiana. Other regions of India also grew wheat, but farmers there lacked assured irrigation and yields were low. In the 1960s rainfed agriculture accounted for 80 percent of cultivated land and about two-thirds of the wheat-growing region.[43] For farmers dependent on rainfall, the timing of rain affected not only their overall yield but also whether it was economic to apply fertilizer.[44] Thus, even before Borlaug became involved, American foundations had laid the groundwork for focusing on irrigated areas, even though this went against the prevailing political goal of social equity.

REORGANIZING THE INDIAN WHEAT PROGRAM

With the help of the RF, Indian agricultural policymakers and prominent scientists began to centralize crop research programs in the 1950s. Wellhausen and Grant noted that decentralized research centers rarely coordinated their work, which impeded progress in maize breeding.[45] This was later confirmed by an agricultural review team consisting of both RF and Indian scientists.[46] In 1956 the Indian government invited the RF to coordinate maize, millet, and sorghum research. The next year the government of India started the Coordinated Maize Breeding Scheme, led by the RF scientists Cummings and Grant. A subcommittee of ICAR's botany division, led by Pal, "recommended the division of the country into . . . four zones for purposes of maize breeding work" in the RF's research program.[47] This idea for breeding according to broad agroclimatic zones was already extant in other rice research programs in Asia, but this was the first time it was applied to India.[48]

The government of India waited over a decade to invite the RF to work on wheat in India. This was, presumably, because Indian researchers felt that the national wheat program was making adequate progress. India's many wheat breeders were releasing popular varieties, but efforts were uncoordinated. Under Pal's influence, ICAR started an informal coordinated wheat program in 1961 "modeled on the coordinated maize program" and put scientists at IARI in Delhi in charge.[49] The RF was not directly involved, but RF scientists guided IARI research away from secondary and ornamental crops and toward wheat.[50] The RF also supported scholarships for wheat scientists to study in the United States and worked closely with Pal to train a generation of wheat researchers.[51]

ICAR first invited Borlaug to consult on wheat research in India in 1963 at the suggestion of M. S. Swaminathan, a cytogeneticist at IARI. Swaminathan had learned of Borlaug through the International Spring

Wheat Yield Nursery, where he noticed Borlaug's semidwarf wheats.[52] Like the tall Mexican wheats, traditional Indian wheats could withstand only so much fertilizer before the grain became too heavy and the plant lodged. On Borlaug's tour of India in 1963 he immediately noticed the lack of fertilizer and its impact on wheat yields. At that point he was focused on wheat breeding for high fertility in Mexico. Borlaug presented his recommendations to ICAR in 1964. He wrote that despite past research efforts focused on local soil fertility levels, "new types of wheat varieties are urgently needed" to survive the "anticipated changes that will come about through the use of heavy rates of chemical fertilizers."[53] Next, he suggested, "the major emphasis for the next 5 to 7 years in breeding should be on improvement of the varieties for irrigated wheat production" to rapidly increase food production in India.[54] All his recommendations were implemented. Borlaug recruited his friend R. Glenn Anderson, a Canadian wheat scientist, to join the RF's India team in late 1964 and to spearhead the new "unified and aggressive" coordinated wheat program.[55] In 1965 the All India Coordinated Wheat Improvement Program was officially launched. This program separated India into five agroclimatic zones for wheat. Anderson became joint coordinator of the All India Coordinated Wheat Improvement Program with S. P. Kohli, an Indian geneticist, as coordinator.[56]

US president Lyndon Johnson was deeply involved in Indian agricultural politics at this time. Instead of the multiyear food aid contracts employed by his predecessors, Johnson strategically used PL-480 as a tool to push his political objectives in India. He used shorter-term contracts and threatened to withhold aid if India did not take certain actions.[57] Around 1964, Johnson and his advisors pressured the Indian government to move toward fertilizer-intensive agriculture by importing and producing chemical fertilizers.[58] India thus far had not focused on producing fertilizers due to concerns about such production causing inequality (for example, one state getting a fertilizer factory and others not).[59] The Indian government also discouraged foreign investment in fertilizer production in India through policies that required companies to have majority domestic ownership.[60] India stuck with these policies until December 1965.

Scholars recognize 1965–1966 as a pivot point in Indian agriculture owing to both political and climatic factors.[61] In this period, President Johnson and his national security advisers used the PL-480 program to bargain with India and catalyze a shift from food aid to self-sufficiency in grains.[62] Simultaneously, India faced a war with Pakistan and droughts and floods. Adding to India's political turmoil was the sudden death of Prime Minister Nehru on May 27, 1964. Johnson capitalized on these crises to push for India's agricultural reform.[63] Patel called these events "a political-ecological 'shock doctrine.'"[64]

Congress named Lal Bahadur Shastri the next prime minister, and Shastri was more supportive of agricultural reform than his predecessor. Shastri made food production a high priority, and appointed Chidambaram Subramaniam as the minister of agriculture. Subramaniam supported investment in commercial agriculture and oversaw reform of the Indian agricultural research system. Subramaniam appointed Pal as ICAR's director general (ICAR's first scientist, rather than bureaucrat, leader), appointed Swaminathan as director of IARI, and put IARI and all agricultural research under the jurisdiction of ICAR. Subramaniam's initiatives were called the "new agricultural strategy" and focused on deploying technology to progressive farmers. Subramaniam stated: "If we concentrate our efforts in a given area where we have assured water supply and we have the necessary extension services also concentrated in that area . . . then it should be possible for us to achieve much better results than by merely dispersing our effort in a thin way throughout the country."[65] This "betting on the strong" approach replaced the earlier focus on social equity.

During Shastri's term, the Indian government implemented a minimum support price for wheat to incentivize production.[66] India also created the Food Corporation of India, which procures and distributes food at subsidized prices. And despite longstanding opposition from the Planning Commission, in December 1965 India's congress agreed to import large quantities of fertilizer and allow foreign investment in fertilizer plants.[67] Kapil Subramanian shows that this change followed a secret meeting, the so-called Treaty of Rome, in November 1965 between Subramaniam and Orville Freeman, secretary of the US Department of Agriculture.[68] These two officials signed a "secret treaty" agreeing that the government of India would increase production and distribution of fertilizer and improve agricultural credit, and that "the new fertilizer responsive varieties of food grains will be planted on well irrigated land, applying from 100 to 150 pounds of fertilizer per acre as compared with a national average of 3 to 5 pounds per acre. These new varieties, planted on the best irrigated land, would get the necessary fertilizer even though this might require a cutback on some other land if fertilizer were in short supply."[69] In 1965 the Indian government shipped 250 tons of Borlaug's Mexican wheat seeds to India for direct planting in irrigated areas, and in 1966 imported 18,000 tons of seeds from Mexico.[70] These quantities were unprecedented. The varieties would be planted in the Intensive Agricultural District Programme areas that were primed for farmer education and demonstrations to the public.[71]

In the new political environment Borlaug and Anderson were well positioned to aggressively pursue an Indian wheat program focused on wide adaptation and higher levels of fertilizer. Borlaug argued that in India, "every effort should be made to develop widely adapted varieties," and further, he aimed to "convince the research scientists that adaptation of wheat varieties seldom or never coincides with state boundaries."[72] Borlaug was also convinced that more fertilizer would shock farmers and scientists into action.[73] He advocated to concentrate fertilizers in the irrigated, high-production areas of India.

Indian scientists followed Borlaug's lead on fertilizers. Borlaug and Swaminathan both agreed on the "betting on the strong" strategy to test varieties on large, irrigated plots.[74] Starting in 1963 the Mexican varieties Sonora 63, Sonora 64, Mayo 64, and Lerma Rojo 64 were tested in India on irrigated, "heavily fertilized land."[75] ICAR also reoriented wheat-breeding programs toward higher levels of fertilizer. ICAR's proposal for the coordinated wheat program stated, "The main emphasis will be placed on breeding varieties for high fertility anticipating that fertilizer will become increasingly available" and "application of heavy rates of nitrogen on varieties bred for irrigation."[76] The wheat project coordinator, Kohli, advocated "breeding for high fertility conditions," a departure from earlier breeding goals.[77] High-ranking scientists such as Kohli, Pal, and Swaminathan worked in close contact with both Borlaug and Anderson to develop a research program that would produce wheats adapted to high-fertility conditions.

The coordinated wheat program centralized power of IARI, India's main agricultural research body, and ICAR, the bureaucratic body, in Delhi.[78] The centralization of control is common to other modernization-era reforms and not unique to India. But IARI scientists had to prove that their research program could still benefit farmers in most of the country. Not everyone was happy with the centralized wheat program, as some scientists raised "serious questions on the part of the states as to whether the Coordinator can fairly represent them," while others noted that "IARI has deliberately usurped credit due to these all-India coordinated projects, to the detriment of the non-IARI staff concerned."[79] Under the auspices of IARI coordination, the centralized research system benefited scientists at the top of IARI's hierarchy. These scientists were the main advocates for wide adaptation.

ADVOCATES OF WIDE ADAPTATION IN INDIA

Even prior to the RF's involvement in India, Indian wheat breeders were part of a global network of international scientists. In the late 1950s Indian plant breeders were interested in theoretical questions of genetics, genotype and environment interactions, plant adaptation, and other contemporary topics in plant breeding.[80] But before the Green Revolution there was no unified approach to breeding under favorable environments and for wide adaptation. The structural shift in the wheat program paralleled an ideological shift toward new breeding goals, both under the RF's advisement.

India's coordinated wheat program did not initially aim to produce widely adapted crop varieties, but wide adaptation became mainstream while the program matured. Wide adaptation happened to support the scientific and administrative goals of the coordinated wheat program, such as centralizing the research headquarters in northwest India. Almost all arguments in favor of wide adaptation came from RF scientists and Indian scientists who worked in northwest India. These scientists had the most political capital as well as personal motivations for promoting wide adaptation as the dominant plant-breeding ideal. They employed wide adaptation as a rhetorical strategy to justify centralizing wheat research around New Delhi.

India's wheat program began to focus on wide adaptation during the involvement of Borlaug and the RF in the 1960s. To start, Indian scientists recognized that day-length sensitivity had limited previous foreign varieties, but "because of the introduction of genetic factors conferring an insensitivity to length of day into the make-up of our varieties, old ideas and the adaptation of varieties had to be given up."[81] Wheat research in India soon focused on how to produce and identify widely adapted varieties of wheat (for example, using summer nurseries for breeding and testing), as well as the biochemical and morphological characteristics of widely adapted varieties. At annual Indian agricultural research meetings in 1966, 1967, and 1969, topics included "genetics of wide adaptation in wheat" and "breeding for wide adaptability."[82] Kohli stated that "to be successful under the Indian crop cultivation conditions, the wheat varieties must be adapted to a wide range of environmental conditions."[83] In addition to importing Borlaug's widely adapted varieties, Indian scientists produced their own evidence to support wide adaptation.

The concept of wide adaptation helped scientists appeal to social critics while simultaneously centralizing research and focusing on irrigated and highly fertilized conditions. Indian scientists used explicitly scientific arguments about wide adaptation to support breeding for favorable environments. Many high-ranking Indian agricultural scientists argued that selecting under optimum environments could lead to varieties with wide adaptation, even in marginal and drought-prone environments. Indian scientists produced their own studies to support this position. They also relied on international literature from American and Japanese researchers (especially the work of Kenneth J. Frey, of Iowa State University, and Kanji Gotoh and Shun-Ichi Osani, of Hokkaido University in Japan) to argue that "a crop plant which shows adaptive response under non-stress environments could also be able to produce adaptive phenotype under stress environments."[84] In other words, a variety bred under irrigated conditions, if widely adapted, could still adapt to drought conditions.

Indian scientists used wide adaptation to argue that a concentrated research program could still benefit average farmers. Swaminathan wrote, "When a variety is evolved under high fertility conditions, it can also yield well when grown under moderate fertility conditions. It is, however, not possible to achieve the reverse."[85] P. N. Bahl, of IARI's Department of Genetics, argued in 1966: "I am highly convinced that all selection work should be done under highly fertile conditions. We should not dissipate our energies by having separate programmes for different kinds of fertility levels. A variety that gives high yield under high fertility and adequate moisture conditions usually proves to be also good under low or average fertility."[86] Bahl and other scientists in India argued that breeding under high fertilizer and irrigation could result in varieties also suited for marginal conditions. An IARI report stated that "selection under favourable conditions appears to be of promise both in breeding for drought resistance and in breeding for wide adaptability."[87] High-profile Indian scientists justified research in favorable conditions through the questionable premise that Borlaug's widely adapted wheats would perform equally as well as if not better than local varieties under less-favorable conditions.

Published articles from 1960 through 1970 show how Indian scientists justified wide adaptation and selection under favorable environments in biological terms. For example, IARI scientists N. N. Roy and B. R. Murty, both of whom studied at Cornell University, produced a study that found varieties selected under high fertility usually had a higher yield than varieties selected under "suboptimal conditions."[88] They reasoned that under suboptimal conditions, environmental variation caused too much phenotypic variability, making selection inefficient. Plant scientists still generally agree with this statement, although some scientists argue the opposite, that "the choice of the selection environment directly determines the potential genetic gains in the target environment. Ideally, the selection environment should mimic the target environment in all aspects."[89]

In a 1970 paper also by Roy and Murty, the authors repeated the argument of RF scientists: "Krull et al. (1966) reported that wide adaptation is genetically controlled and that varieties do not require to be bred for the particular conditions."[90] Roy and Murty asserted that because wide adaptation is genetically inherited, varieties selected for their yield response under high fertility and irrigated conditions would still produce high yields in marginal conditions. At a 1967 conference, Roy and R. D. Asana, also from IARI, gave a different biological reason for wide adaptation. They argued, "Although it might be possible eventually to define the characteristics for the ideal plants adapted to a specific environment like drought, it would be much more difficult to define all possible combinations of a range of characteristics necessary to provide good adaptability in an otherwise fluctuating environment."[91] Roy and Asana implicitly argued that it was better to have a variety that could adapt to a range of conditions than one that was adapted to one specific condition.

To prove the wide adaptation of a variety, Indian scientists tested new wheat varieties under different locations and agronomic conditions. Only the varieties with high average yields would be approved for release to seed companies and farmers. While most breeding was done under favorable environments, the program coordinators promised that "the extensive breeding material which will become available from the strengthened programs at the main centres will be fully tested under *barani* [rainfed] conditions and the most promising strains will be selected for release to cultivators for *barani* production" and that "a variety performing well under both conditions would be desirable."[92] Thus, scientists at the central research facilities argued that they could provide varieties widely adapted across agroclimatic zones and conditions.

Many of the scientists who argued for wide adaptation and selection under favorable environments were affiliated with IARI. As previously noted, IARI served as a hub for the Indian coordinated wheat program and benefited from the centralization of power at its New Delhi headquarters. However, outside of the IARI research stations, theory did not match practice in most of the country.

QUESTIONING WIDE ADAPTATION AND THE NEW AGRICULTURAL STRATEGY

Borlaug's Mexican wheats were approved for release in India based on results of international yield trials and claims of their wide adaptation.[93] The Mexican variety Lerma Rojo 64 was especially celebrated for its "adaptability over wide regions."[94] Not all Indian scientists agreed with these claims. Some wondered whether the wheat program, focused on irrigated agriculture in the northwest, would also serve the needs of the extensive area of rainfed agriculture. Adding to the controversy, the Mexican varieties appeared to flourish only under high-fertility conditions. The first batch of Mexican varieties grew poorly due to damage and defects to the seeds, which were hastily prepared by the Mexican government. Later seed batches had higher yields, but in 1965, after two years of testing, the Mexican varieties still did not perform as well as expected. Scientists realized that this was attributable to their different agronomic needs from the tall Indian varieties, such as the need for irrigation and a shallower planting depth. The Mexican wheats required irrigation at more precise times, which was not usually possible under prevailing canal irrigation systems in India. Borlaug and his colleagues had introduced the semidwarf wheat into the commercial irrigated areas of Mexico, but India was a different landscape. At the time of the Green Revolution about two-thirds of wheat was not irrigated, and even within irrigated areas the quality of irrigation varied.

Scientists affiliated with the Ford and Rockefeller Foundations noted the lack of a steady water supply even in areas classified as irrigated. Anderson wrote to Borlaug in 1966: "We have as yet not had any winter rains in the Delhi area and the crops which are not under irrigation are very thin and I am afraid yields will be quite low. Even on the irrigated acreage that is under canal control, many of the crops are suffering appreciably from lack of water because the canals have not been delivering water at regular intervals."[95] Indeed, many of India's canals "were designed to provide protection against drought rather than productive irrigation."[96] Cummings, upon visiting a tube well project in Uttar Pradesh, wrote, "The technology in establishing and maintaining the wells was extremely poor," with many "power failures and actual burning out."[97] Lack of good irrigation was a major challenge to implementing the Mexican varieties in India. Yet most scientific discussions after 1965 centered not on whether the Mexican varieties should be promoted but rather on the amount of fertilizer to be used in field experiments. Anderson argued that field trials did not have high enough levels of fertilizer, and that "the Mexican varieties show their potential only under very high fertility levels."[98] This of course directly contradicts Borlaug's dogma that the Mexican varieties outyield all other varieties, fertilized or unfertilized.

In response to the failure of Mexican wheats in the early coordinated trials, a "committee on high fertility and agronomic trials" from the Indian coordinated wheat program decided that for the next season, the Mexican varieties should be tested in separate trials with higher levels of fertilizer.[99] In fact, the committee decided in 1967 that these Uniform High Fertility Trials would test *only* at 135 kilograms of nitrogenous fertilizer per hectare (120 pounds per acre), the level that Borlaug recommended for semidwarf wheat in India.[100] Meanwhile, the average rate of fertilizers applied to cropland in India in 1970 was 11 kilograms per hectare (kg/ha), although this rate was higher and rapidly increasing for irrigated wheat.[101] Only in recent years has the average fertilizer use in India even approached 135 kilograms of nitrogenous fertilizer per hectare.

The initial poor performance of the Mexican varieties in India concerned both scientists and economists, especially the economists in India's Planning Commission. David Hopper, of the Ford Foundation in India, wrote to Borlaug: "The Mexican developed varieties gave only marginally better responses to nitrogen application, and on the basis of these results the Planning Commission is raising questions about the policy of the Agricultural Ministry that seeks to concentrate nitrogen, which is in short supply, primarily in the areas where the exotic varieties will be grown."[102] Further, one Planning Commission member asked "whether the Indian varieties which showed high response to fertilization at the lower levels may be fully exploited instead of using heavy doses of the limited quantity of the fertilizer with the Mexican varieties . . . especially when there is a shortage of fertilizers in the country."[103] The Indian economic planners believed that fertilizer should be distributed equally throughout the country, which Borlaug and Anderson fervently resisted.[104]

The new agricultural strategy outraged many economists. In 1966 economists B. S. Minhas and T. N. Srinivasan wrote an article in the Planning Commission's official publication, *Yojana*. They argued that instead of the strategy that concentrated fertilizers on irrigated areas planted with new varieties, fertilizer should be spread more evenly between old and new varieties.[105] This echoed the Planning Commission's equity-based approach to agricultural development but added a quantitative analysis of varietal responses to fertilizer. Minhas and Srinivasan explained that both local and imported wheat varieties had strong yield responses to low levels of fertilizer. Therefore, the country could achieve higher crop production if fertilizer was dispersed at lower levels across the country rather than applied at near the maximum dose to only the new varieties, because fertilizers have diminishing returns at higher levels. While the imported wheat and rice varieties could respond to much higher levels of fertilizer than local varieties could, they argued that the concentration approach was not feasible, and that "the extremely high dosages of chemical nutrients recommended for the new varieties" were not supported by the existing experimental data.[106] Subramanian summarized the article as follows: "While many might have questioned the policy of concentrating resources from the point of view of equity, Minhas and Srinivasan questioned it from the perspective of production, on the altar of which long standing ideas about equity were being ostensibly sacrificed."[107]

In private correspondence, some RF scientists in India agreed that spreading out fertilizer would improve production overall but would not provoke the required change in mentality among Indian farmers or scientists. The Indian plant scientist Vinayak Govind Panse had questioned the high dosage of fertilizers for the coordinated maize program. RF agronomist Bill C. Wright wrote back, "It is unquestionably true that if one wishes to obtain the maximum response of grain for each kilogram of nitrogen fertilizer used, a small dose should be applied to many hectares rather than applying larger doses to a few hectares."[108] But he asked further, "Where does one stop?" and noted that a better strategy would be to concentrate fertilizer in the hands of better farmers.[109]

Anderson's notes and reports from the early 1970s show that the high rates of fertilizer were not necessarily economical but meant to convert farmers to using chemical fertilizers. Anderson stated in a 1972 speech, "A high rate of application must be recommended and used to maximize yield for purposes of demonstration effect. One should not be concerned with economics at this point. This initial demonstration will convince farmers, extension workers, legislators and planners if they are shown the results."[110] He wrote a year later, "The economist said if you put 30 pounds of fertilizer on 3x acres, you would increase production to a higher level than if you applied 120 pounds on x acres. This was true. On the biological side, we argued that the latter condition should be followed because of the psychological shock effect on the farmer."[111] Thus, Anderson actually admitted that the economists were correct.

This idea of psychological shock pervaded Borlaug and Anderson's discourse on fertilizers. Borlaug felt that Indian scientists were too academic and needed to be shocked out of complacency by seeing the semidwarf varieties grown with high fertilizer levels. While Borlaug and Anderson were not trained in economics or psychology, Jonathan Harwood explained that "they were merely echoing what proponents of modernisation theory had been saying for some time: that psychological and cultural change were just as significant in the modernisation process as social and economic change."[112] Harwood also asked, if maximizing food production was not the goal of the Green Revolution, then what was? He agrees with Harry M. Cleaver Jr.'s 1972 analysis that the aim of the Green Revolution was not to increase food production but to commercialize agriculture as a strategy of modernization.[113]

These debates over fertilizer threatened the entire premise of the RF's Indian wheat program. In a 1966 RF field directors' meeting, Borlaug was enraged, stating, "There is a complete lack of understanding among people in government of the kinds and quantities of fertilizer needed."[114] Borlaug wrote in 1966, "I am against this dispersion and dilution of fer-

tilizer application, and instead feel that the program should try to produce tremendous yield increases on the area where the dwarf varieties
can be heavily fertilized and properly watered."[115] Borlaug felt there was
no alternative solution to the food problem. He wrote, "India must really
think big and positively on fertilizer, or starve."[116]

Indian scientists also held opposing opinions over the strategy of selecting wheat in ideal environments. At least one strong argument exists
from J. S. Kanwar, a respected soil scientist and the deputy director general of ICAR at that time. Kanwar reacted against the push for the Mexican varieties on a wide scale as well as breeding and testing under only
highly fertilized, irrigated, and managed conditions. At a 1969 wheat
research meeting, Kanwar stated: "It will be drought tolerant or drought
escaping varieties that are badly needed for different trials. It is suggested
that the breeding programme for wheat be oriented to develop varieties
for rainfed conditions. An active testing and demonstration programme
is also required. . . . There is a need for an active programme of selection
of varieties for rainfed conditions."[117] Kanwar, while not explicitly dis-
agreeing with wide adaptation, highlighted that "there is very inadequate
data about the adaptability of new high yielding varieties for rainfed and
unirrigated conditions."[118] Kanwar and a few other scientists resisted the
new regime of research and held to the more traditional viewpoint that
crops should be bred in the conditions that they are grown.

Even Cummings, the director of the RF's Indian Agriculture Program, was skeptical of how easily the Mexican wheats would perform in
marginal areas, stating that "as a complementary variety, the best Indian
variety should be recommended" in 1965.[119] Cummings wrote in his diary that the RF should be cautious about releasing the Mexican varieties:
"One would need to appraise carefully the performance of these wheats
this year, and the desirability of going reasonably cautiously until one
had had more experience in farmer tests and certainly one would want
to consider utilizing them only in situations where there are ample irrigation facilities and where farmers are in a position to have available
and utilize large quantities of fertilizer."[120] Cummings's hesitation likely
stemmed from his experience with hybrid maize in India, which was
poorly adapted to Indian conditions. He was also concerned about the
availability of irrigation and fertilizers. But Borlaug was still convinced
that the Mexican wheat agenda should move ahead as planned. In 1967
Borlaug encouraged his collaborators in India and Pakistan to "*Get them*
[the seeds] *multiplied—abandon the three year yield testing program that
is being followed.*"[121] Borlaug did not waver from his plan of releasing the
Mexican varieties on a massive scale in South Asia.

Other Indian scientists began to note that the Mexican wheats did not live up to their reputation as universally high yielding. At a 1967 conference S. M. Gandhi, a wheat scientist working in the mostly dryland state of Rajasthan, reported: "Yield results obtained under rainfed conditions are particularly interesting in the present context of intensive efforts in breeding short strawed varieties. C306, CA82, K65 and D144, all tall [and Indian] varieties, have come in the first group of highest yielding entries in the series under rainfed conditions."[122] However, he also noted that the tall varieties did not perform well under high-fertility conditions and intensive irrigation. In other words, the "Mexican wheats were superior under intensive farming conditions while, Indian varieties were superior under average or below average conditions."[123] Two years later Kanwar reported that "it is also not possible to conclude anything regarding the performance of high yielding varieties as different varieties including local variety were not compared. More trials are needed."[124] Kanwar is correct that the wheat yield trials typically only tested new varieties against major commercial varieties, leaving out local varieties. Meanwhile, the RF's Wright noted that in the state of Gujarat, "tall varieties have consistently yielded as good or even better than the dwarf wheats," though he considered this "unusual."[125]

Then in 1967 Indian scientists decided to not include the Mexican varieties in the southern peninsular zone experiments because the Indian varieties had consistently higher yields. This prompted one Indian scientist to argue, "I do not agree with Dr. Upadhyaya that dwarf varieties have no place in the Peninsular Zone if proper agronomy is practiced."[126] In response, Y. M. Upadhyaya, who was working for IARI in Indore, Madhya Pradesh, simply stated, "Proper agronomy is not possible under farmers' conditions."[127] The widely adapted wheats demanded "proper" use of fertilizers and irrigation, along with "proper" preparation of fields and timely harvesting. As ICAR's Kanwar succinctly stated, "Miracle seeds do not produce miracles unless the right combination of factors such as fertilisers, soil management, water management and crop management practices interact."[128]

QUANTIFYING LERMA ROJO 64'S ADAPTATION

Despite the setbacks, Borlaug and Anderson fiercely promoted the release of the Mexican semidwarf variety Lerma Rojo 64 in India based on its high yields and wide adaptation. India imported Lerma Rojo 64 in large quantities in the 1960s and it was one of the first Mexican varieties approved in India. Borlaug himself stated in his 1967 oral history that Lerma Rojo 64's most important trait was its "wide adaptation—fertilized or unfertilized."[129] Anderson recalled at a 1969 conference in India, "I am firmly convinced that the adoption of cultivation of dwarf wheats in this

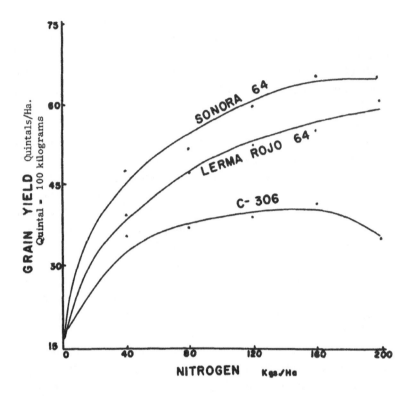

FIGURE 2.1. Plot of the yield response to increasing levels of fertilizers for the wheat varieties Sonora 64 (Mexican, semidwarf), Lerma Rojo 64 (Mexican, semidwarf), and C306 (Indian, tall). Image reproduced from United States Department of Agriculture, *Biology of Rust Resistance in Forest Trees: Proceedings of a NATO-IUFRO Advanced Study Institute, August 17–24, 1969* (Washington, DC: US Forest Service, 1972). Original caption: "Differential nitrogen response curves for two Mexican dwarf wheat varieties compared to one of the best tall-strawed Indian varieties C 306, at the Uttar Pradesh Agricultural University, Pantnagar, U.P., India in 1966. Data by Drs. K C. Sharma, D. Misra and B. C. Wright."

country was materially hastened through the wide adaptation of Lerma Rojo."[130] For a few years in the 1960s Lerma Rojo 64 gave outstanding yields in the RF's international wheat yield trials, which Borlaug took as proof of its wide adaptation. Historical evidence, however, suggests that Lerma Rojo 64 is adapted to specific conditions of high fertility but was able to be grown over a wide area owing to its photoperiod insensitivity and its positive response to fertilizers.

Borlaug frequently used the graph included here as figure 2.1 to support the superiority of Lerma Rojo 64 to tall Indian varieties such as the recently released C306 (which was widely adopted in rainfed parts of India). This figure shows the yield of three varieties—Sonora 64, Lerma Rojo 64, and C306—grown under fertilizer levels ranging from 0 to 200 kg/ha. It shows that C306 responds less to higher fertilizer rates, while the two Mexican varieties have higher yield responses at these rates. The data for the figure were based on an agronomic test performed in 1966 by K. C. Sharma and D. Misra, of Pantnagar, and Wright, of the RF. A version of this figure appeared in the 1965–1966 RF Agricultural Sciences Program Annual Report, and Borlaug included it in various letters, publications, and talks about the RF's Indian Agricultural Program.

Fertilizer studies of Mexican and Indian varieties by different scientists yielded different results, especially under low-fertilizer conditions. K. S. Gill's book *Research on Dwarf Wheats* (1979), as cited by Subramanian, compared the average yields at different fertilizer levels of tall Indian varieties, the Mexican Sonora 64 and Lerma Rojo 64 varieties, and semidwarf varieties bred in India.[131] Subramanian wrote: "These showed that with no fertilizer applied, the tall varieties did better than the dwarfs. At low dosages, they produced higher yields than dwarf varieties; an advantage they lost somewhere between 40 and 80 kg of nitrogen per hectare. Yet, even at the very high level of 200 kg of nitrogen/ ha, their yield was only 20% lower than the best dwarf varieties."[132] An analysis by Kohli also found that C306 outyielded the Mexican and Mexican-derived varieties in lower-fertility environments, but the opposite was true in high-fertility environments.[133] On the other hand, Minhas and Srinivasan found that the Mexican wheats "give higher yields than existing varieties even without the use of fertilisers," thus agreeing with the findings shown in figure 2.1.[134] Kirit S. Parikh used data from farmer field experiments in the late 1960s and found that "the HYV does dominate the local variety even at zero fertiliser level."[135] It seems that as with many aspects of agriculture, the results depend on the context of the experiment and the method of analysis.

There is evidence, however, that Lerma Rojo 64 was not as widely adapted as Borlaug claimed. Keith Finlay wrote to Borlaug in 1966, "Lerma Rojo 64 is of particular interest because it is even less stable—indicating that it is better adapted to 'higher yielding' environments."[136] Finlay applied his yield stability model to Borlaug's data from the international wheat trials. While he found that the Mexican varieties were generally higher yielding and more stable across locations, the yields varied by year. Borlaug had responded earlier to Finlay's concern about a different semidwarf, Sonora 64, and its low yields: "In trying to develop as rapidly as possible a dwarf variety with good gluten quality, we settled for a variety, such as Sonora 64, with less yield stability. This variety is high

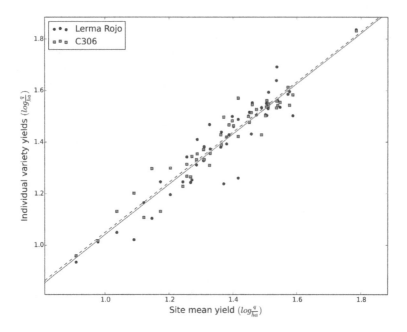

FIGURE 2.2. Regression analysis of varieties C306 (solid line) and Lerma Rojo 64 (dashed line) for 1964–65 and 1965–66 wheat seasons in northwest India, over both rainfed and irrigated conditions. Lerma Rojo's average yield was 27.3 ± 11.2 quintals per hectare and C306 was 27.5 ± 10.0 quintals per hectare. Figure rendered by Erick Peirson.

yielding when grown under heavy fertilizer conditions and is properly irrigated . . . but it will produce lower yields generally when in the hands of the average farmer."[137] This statement is quite surprising, as it goes completely against nearly all his public statements about the Mexican varieties.

Considering the confusion over Lerma Rojo 64's wide adaptation, I used Finlay and Wilkinson's analysis to test the adaptation of Lerma Rojo 64 and the popular tall Indian variety C306 using yield trial data from India (fig. 2.2). Data were collected from the *Results of the Coordinated Wheat Trial Results*, 1964–1965 and 1965–1966 for the Northwest Plain Zone Uniform Regional Trials. The varieties were tested under a wide range of locations and conditions that included irrigated or non-irrigated and nitrogen fertilizer levels that ranged from 45 to 90 kg/ha in 1964–1965 and in the range of 0 to 120 kg/ha in 1965–1966. Unlike Wright's figure that plotted only yield of a variety by fertilizer level, Finlay's model is a logarithmic plot of the yield of a variety over the average yield at that location (a proxy for environmental condition). A variety

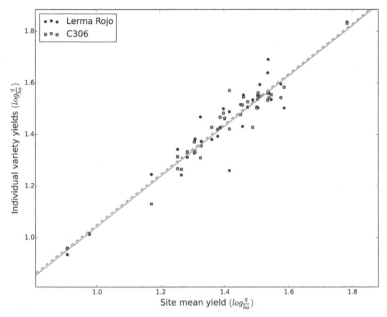

FIGURE 2.3. Regression analysis of varieties C306 (solid line) and Lerma Rojo 64 (dashed line) for 1964–65 and 1965–66 wheat seasons in northwest India, over only irrigated conditions. Figure rendered by Erick Peirson.

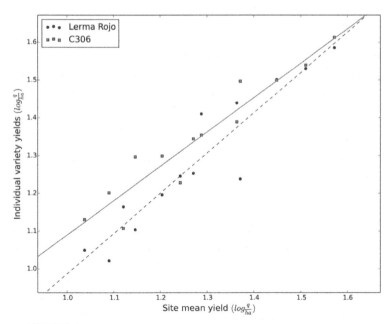

FIGURE 2.4. Regression analysis of varieties C306 (solid line) and Lerma Rojo 64 (dashed line) for 1964–65 and 1965–66 wheat seasons in northwest India, over only rainfed conditions. Figure rendered by Erick Peirson.

TABLE 2.1. Results of regression analysis of the varieties C306 and Lerma Rojo 64.

	Both conditions	Irrigated only	Rainfed only
C306 slope	0.94	0.98	0.91
Lerma Rojo 64 slope	1.02	0.97	1.07
C306 average yield (q/ha)	27.5	28.9	23.6
Lerma Rojo 64 average yield (q/ha)	27.2	29.6	21.1

with a high regression coefficient (slope) is better adapted to favorable environments, and a low slope indicates stability across conditions, or wide adaptation. A slope of 1 is average stability.

According to Finlay's model, C306 and Lerma Rojo 64 did not have significantly different yield responses. Table 2.1 shows the results of my analysis, including the slope and average yield (in quintals per hectare or q/ha; a quintal is 100 kilograms) of each variety under irrigated, rainfed, and both conditions combined. Both varieties show average stability (a slope near 1), though C306 is more stable (slope of 0.936) and Lerma Rojo 64 is slightly more adapted to better environments (slope of 1.024) (table 2.1). This counters Borlaug's argument that the Mexican varieties were more widely adapted than local varieties. This is partly because Borlaug and Finlay appealed to different definitions of wide adaptation, with Finlay focused more on environmental conditions and Borlaug focused more on location and yield. Further, if I follow Finlay's 1966 analysis of Borlaug's data and separate the data into irrigated and rainfed conditions, the results change noticeably (fig. 2.3, fig. 2.4). While both varieties performed about equally under irrigated conditions, under rainfed conditions C306 was even more stable and higher yielding than Lerma Rojo 64.

My analysis was limited to the data available from the 1964–1965 and 1965–1966 in northwest India, which were the only available data that included the Mexican-derived varieties in India over a variety of conditions. Nonetheless, my analysis shows that Lerma Rojo 64 is better adapted to irrigated conditions and it more strongly suggests that C306, the tall Indian variety, is not only widely adapted in India but had more stable (and higher) yields under rainfed conditions. Even under irrigated conditions and varying levels of fertilizer, Lerma Rojo 64 did not have higher average yields than C306. Trials conducted at even higher fertilizer levels could have different outcomes, given Lerma Rojo 64's response to fertilizer. My results support the finding that the semidwarf varieties performed well under trial conditions, but worse under average farm

conditions. These results refute those who would ask whether food production could have increased without the Green Revolution. I agree with
Subramanian's argument that the "transformation of wheat production
could have been achieved (and more cheaply) without the dwarf varieties
altogether."[138]

My analysis supports B. H. Farmer's evaluation of Green Revolution
rice varieties in South Asia.[139] Farmer wrote that Green Revolution varieties were photoperiod insensitive, thus making them adapted to a broad
range of locations under adequate agronomic conditions. RF-affiliated
scientists, however, claimed a second type of adaptation: adaptation
across varied environments that included pests, monsoon, and drought.
Farmer wrote that "the two kinds of adaptability have been confused, or
at any rate not sufficiently distinguished."[140] This appears to be the case
with Borlaug and the Mexican semidwarf wheats as well, which were
photoperiod insensitive and thus better adapted to Indian conditions
than other foreign wheats. Mexican wheats showed wide adaptability
across locations with similar agroclimatic conditions and management,
but not across field conditions of the average farmer.

Ultimately, scientists in India adopted wide adaptation for a variety
of reasons. Wide adaptation made pragmatic sense to maximize resource
allocation and to bring together decentralized wheat research under one
top-down system. It also allowed wheat breeders to work toward a new
blockbuster wheat variety that would bring them personal and professional prestige. But wide adaptation was based on some questionable assumptions that played out once Borlaug's new varieties were planted on
a wide scale in India. As wide adaptation became codified, it had real
consequences for resource-poor farmers in India.

MYTHBUSTING WIDE ADAPTATION IN INDIA

Over the past fifty years, many scholars and activists have criticized the
impacts of Green Revolution agriculture in India. Some of these critiques focused on the unequal socioeconomic spread of technologies
that favored the larger, irrigated, commercial farms of the Punjab region
over smaller, rainfed farms.[141] Fewer of these critiques identified that the
varieties of wheat and rice released in the mid-1960s were not adapted
to low-rainfall, low-fertility conditions that marginal farmers typically face.[142] But very few critiques have addressed why, in the mid-1960s,
Indian agricultural scientists decided to focus the national system of
wheat research on high-fertility and irrigated conditions. Understanding
how wide adaptation became the dominant framework in Indian plant
breeding is critical to understanding why this paradigm that still exists
in India.

Scientists used wide adaptation to justify radical changes in Indian agricultural policy during a time of political turmoil. Indian and RF scientists spread the new doctrine of plant breeding that was critical to the Green Revolution style of research. While some of these changes were meant as a stopgap for the food crisis, they became embedded as policy and have remained the norm. Wide adaptation became a vehicle to shuttle in new varieties and technologies with the promise that they would promote social equity, while simultaneously cementing a research regime focused on the agroclimatic conditions of northwest India. These changes have led to a systemic bias against improving marginal agriculture in India. Much-needed improvements in irrigation, rural infrastructure, and market access are overshadowed by investments in plant breeding and the release of new varieties.

The story of Borlaug's Mexican wheats in India was supposed to be a triumphant success. India, after all, declared a "wheat revolution" in 1968, and even issued a commemorative stamp. Behind the headlines, however, India's scientists were grappling with ideas of independence, self-sufficiency, social equity, and foreign intervention. Perhaps under different circumstances, Indian wheat scientists would have stood on firmer ground for more thorough vetting of the semidwarf varieties and the Planning Commission would have continued their social-equity-based policies. But India was facing Johnson's strong-arm tactics, which exacerbated Indian scientists' search for a silver bullet solution. National security advisor Robert William "Blowtorch Bob" Komer wrote to Johnson in 1966:

> We finally have the Indians where you've wanted them ever since last April—with the slate wiped clean of previous commitments and India coming to us asking for a new relationship on the terms we want. Circumstances helped (famine and the Pak/Indian war), but seldom has a visit been more carefully prepared, nor the Indians forced more skillfully to come to us. . . . Similarly, you have already proved how our holding back on PL-480 can force India into revolutionizing its agriculture. Once the famine is licked, I'm for continuing to ride PL-480 with a short rein—it will be painful but productive. If these points don't add up to requiring self-help, I'll eat them.[143]

Because of Johnson's pressure by withholding food aid, Indian politicians abandoned their earlier focus on equity and moved instead toward concentrating resources.

This chapter brings up some tough questions about Borlaug. He bypassed the concerns of Indian social planners, scientists, and other international colleagues because he felt the Mexican varieties plus fertilizers were the only solution to India's food problems, despite wheat being a

relatively minor crop in India. Borlaug overlooked other possible pathways and reduced India to a problem of fertilizers and wheat varieties.
I was struck by the phrase "proper agronomy" and Borlaug's statement
that Sonora 64 was "high yielding when grown under heavy fertilizer
conditions and is properly irrigated . . . but it will produce lower yields
generally when in the hands of the average farmer."[144] There is no question that Borlaug knew that at least some of the varieties were not suited to the "average farmer." Unfortunately, the average farmer in India
still does not have the resources, information, or access to apply "proper
agronomy."

It seems that Borlaug promoted a fundamental misinterpretation of
wide adaptation. I argue that Borlaug's widely adapted wheats were so
because of their photoperiod insensitivity and fertilizer responsiveness,
and not because of any genetic or inherent wide adaptation or yield. But
Borlaug fallaciously promoted wide adaptation as adaptation even across
agronomic conditions. Borlaug was focused on a specific set of farmers
who had access to irrigation and fertilizers and used wide adaptation to
deflect criticisms of bias. Regardless of how the story is framed, there are
no easy explanations for why Borlaug held onto the idea of the widely
adapted genotype against both the scientific evidence and the realities of
resource-limited agriculture in India.

CHAPTER 3

INDIAN WHEAT RESEARCH AFTER THE GREEN REVOLUTION

By 1968 India had declared a wheat revolution and US Agency for International Development (USAID) administrator William Gaud proclaimed a Green Revolution in Asia. Indian farmers had adopted new semidwarf wheat varieties at rapid rates and overall wheat production had increased. But soon after 1968, excitement over the semidwarfs faded. Top-level scientists, administrators, and policymakers recognized that the wheat program had mostly benefited large farmers in the Punjab who had access to irrigation and synthetic fertilizers. Questions about the socioeconomic impacts of the Green Revolution were rampant. In the 1970s some of these actors pushed for more research directed at rainfed and marginal agriculture, while others made clear that the large area of dryland wheat in central and southern India had been bypassed by the wheat revolution.

Clearly the Green Revolution was not the miracle some had hoped. While wheat production did increase, the concentrated investment in irrigated areas led to social disparities, even within those regions.[1] Rising energy prices and drought throttled the availability of fertilizer and irrigation.[2] In 1974, wheat scientists noted, "farmers were inclined to revert to local wheats because of input shortages and better prices for the local wheat and straw."[3] There were storage and distribution problems, and a string of good years was succeeded by bad weather in the early 1970s.

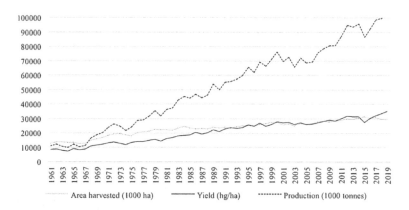

FIGURE 3.1. India's wheat production in 1000 tonnes, wheat area harvested in 1000 hectares (ha), and wheat yield in hectograms per hectare (hg/ha) from 1961 to 2019. Data obtained from Food and Agriculture Organization of the United Nations, *FAOSTAT*.

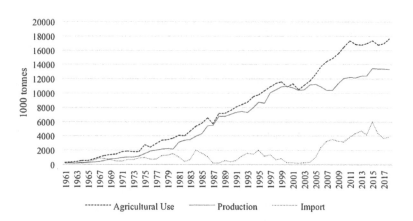

FIGURE 3.2. India's nitrogen fertilizer use, production, and imports in 1000 tonnes from 1961 to 2018. Data obtained from Food and Agriculture Organization of the United Nations, *FAOSTAT*.

Some scholars have suggested that the rapid gains in wheat production from 1966 to 1971 were more likely a result of good weather and higher wheat acreage rather than increased yield from semidwarf wheat.[4] India was once again importing grains in the mid-1970s, and the spread of Green Revolution varieties slowed.

Other than a slight nadir in the 1970s, wheat production in India has continued to rise at a roughly linear rate (fig. 3.1). This corresponds with

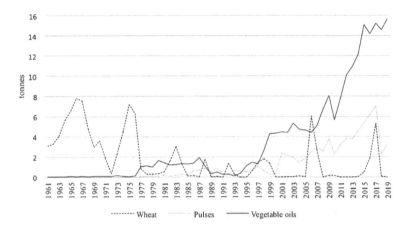

FIGURE 3.3. India's annual imports of wheat, pulses, and vegetable oils in tonnes from 1961 to 2019. Data obtained from Food and Agriculture Organization of the United Nations, *FAOSTAT*.

fertilizer production and consumption that have also increased since the 1960s (fig. 3.2) and expanded irrigation. Yet net food grain availability per capita per year has not increased since the Green Revolution, and per capita availability of protein-rich gram and pulses has declined while wheat availability per capita has increased.[5] Farmers were choosing to grow wheat over these more nutritious crops because of the new price supports. While the US government pressured India to become self-sufficient in wheat and rice, the tradeoff has been that India now relies on imports of pulses and vegetable oils (fig. 3.3).

In the late 1960s the United States sought a more robust understanding of the food situation in India. The USAID commissioned Francine Frankel and the World Bank sent Wolf Ladejinsky to assess the Green Revolution in India.[6] Both scholars released reports in 1969 that strongly critiqued the unequal impacts of the Green Revolution on different socioeconomic classes of farmers.[7] Mainly, these criticisms were that adoption of new varieties was limited mostly to irrigated areas, and that farmers with more capital benefited more from new technologies, despite the fact that many scientists claimed that the seeds and technologies were "scale neutral." In 1975 economist Edward Tenenbaum reported to the World Bank, "Whatever the effects may have been on agricultural productivity, the effects of the green revolution on income distribution have been regressive."[8] Frankel, Ladejinsky, and Tenenbaum and other researchers all pointed out the shortcomings of the Green Revolution despite the increase in wheat yields.[9]

Beyond the socioeconomic impacts of the Green Revolution, food security was also a concern. A 1973 USAID report found that the Green Revolution had not changed the per capita availability of cereal grains in India. Even the Rockefeller Foundation staff in India wrote, "India has made no real progress in improving her people-food equation in the decade of 'the green revolution' and there's no new agricultural technology on the drawing board as glamorous and promising as was the HYV's [high yielding varieties]."[10] The adoption of Green Revolution varieties, technologies, and policy changes increased wheat and rice production, but hunger and malnutrition remain pervasive problems in India. The Green Revolution narrative of preventing mass starvation in India has nevertheless become canon among agricultural scientists and development professionals.

In this chapter I examine Indian wheat research in the wake of the Green Revolution. While the tumultuous mid-1960s led to a change in Indian wheat breeding goals, the post–Green Revolution era brought serious critiques of these changes. Indian scientists also asserted their own project goals and more critically examined assumptions about wide adaptation. Influential factors in this period included new administrators of the coordinated wheat project; new analyses of wheat yield stability; and ongoing failures of Green Revolution wheat varieties—including the "oversuccess" of two Mexican-derived wheat varieties, Kalyan Sona and Sonalika.

EXPECTATIONS MEET REALITY IN THE POST–GREEN REVOLUTION YEARS

In 1967, the Indian Council of Agricultural Research (ICAR)'s Central Variety Release Committee approved the release of two new semidwarf wheat varieties, known as Kalyan Sona and Sonalika. Indian scientists had selected these from lines of the semidwarf wheats from Mexico.[11] Scientists and the government promoted these varieties throughout north India for their yield potential, and farmers rapidly adopted Kalyan Sona and Sonalika under both irrigated and rainfed conditions. These varieties quickly replaced the direct introductions from Mexico—Sonora 64 and Lerma Rojo 64—which both had an undesirable red grain color.

Kalyan Sona and Sonalika dominated wheat production in north India for over a decade.[12] Even the traditionally non-wheat-growing region of northeast India started growing these varieties.[13] Kalyan Sona and Sonalika reigned supreme until the release of several new varieties in the mid- to late 1970s such as HD2009, WL711, and WH147 in the northwest and UP262 in the northeast, but even by the mid-1990s Sonalika was still common in the northeast. Kalyan Sona and Sonalika exemplified Norman Borlaug's research program for wide adaptation because of their high yield and adaptation to both irrigated and rainfed areas. But as one can imagine, the dominance of one or two varieties is not optimal for disease resistance or climate resilience. A monoculture is vulnerable to pests, disease, and abiotic stress. M. V. Rao, the new coordinator of the Indian wheat program, stated in 1975, "Large tracts in the country are occupied by very few varieties primarily Kalyan Sona and Sonalika. This is a dangerous situation and if any new race of any pathogen comes up in severe form, the situation would be disastrous."[14] Fortunately, the varieties were not vulnerable to any major disease at the time, but they were sensitive to climatic stress.

While Kalyan Sona and Sonalika were adopted by farmers at an impressive rate, there were some issues with these varieties. Indian scientists observed that both were thermosensitive (sensitive to temperatures) during their growth stage. This could result in reduced yields and would be an especially significant problem if farmers were not growing other varieties of wheat to offset their losses. Rao stated in 1977, "Varieties like Sonalika, Kalyan Sona suffered the most due to high temperature. On the other hand locally bred or selected materials . . . were least affected."[15] J. P. Srivastava, a pathologist, found similar results and wrote to the RF's R. Glenn Anderson in 1972: "I have data indicating differential response of varieties to higher temps in the early stages of plant growth. The indigenous Indian varieties grow and tiller normally in spite of above normal temperatures. Short duration varieties like Sherbati Sonora, Hira, UP301, UP310, Sonalika and Safed Lerma, at higher temps, completed their vegetative growth rather quickly and entered into reproductive growth. This resulted in reduced tillering, stunted growth, small ears, increased sterility and low yields."[16] Despite the temperature sensitivity of Kalyan Sona and Sonalika, Indian scientists struggled to breed higher-yielding varieties until the CIMMYT-derived variety PBW343 was introduced in the mid-1990s.

Fertilizer supply was also a continuous problem in the 1970s as a result of rising energy prices and the low levels of domestic fertilizer production. Eugene Saari, a CIMMYT staff member in India, wrote to Anderson in 1972: "Fertilizer supplies (N and P) are short by 30%. . . . There is a big push to increase the acreage but it will take about a million hectares to make up the fertilizer deficit."[17] Anderson noted that in 1973, fertilizer supplies were 40 percent short for *rabi* (fall-planted) crops, including wheat.[18] This shortage seriously undermined the potential benefits of the semidwarf wheats.

Though fertilizer consumption in India quadrupled from 1965 to 1974 at "an almost unparalleled rate in the history of world agriculture," the average application rate in 1970–1971 was still only 12 kilograms per hectare (kg/ha) and varied widely between states.[19] Farmers applied more fertilizer to wheat, around 51 kilograms of nitrogen per hectare (kg N/ha) in 1970–1971, with the expectation of higher yields from the new varieties.[20] This rate of fertilizer application was still less than half the official state recommended dosages, which were about 120 kg N/ha plus smaller quantities of phosphorus and potassium.[21]

The World Bank contracted economist Edward Tenenbaum to report on the fertilizer market situation in India. Tenenbaum reported that farmers were using lower than the recommended levels because they were maximizing their profit. He wrote that while earlier fertilizer trials showed linear yield increases at high fertilizer rates, this result was due to the sandy, low fertility soils in the Punjab region where many of the experiments were conducted. An average farmer in India faced diminishing yield returns at higher fertilizer rates. He found that even in Punjab, "the use of new 'High Yielding Varieties' increased almost 20 times, the frequency of irrigation increased 17%, and use of fertilizer multiplied by 4.6. Yet yields per hectare were only around 10% greater, and have been falling fairly steadily from a peak reached in 1968–69. Finally, average yields per kilogram of fertilizer have declined dramatically."[22] Banks also found the official fertilizer rates too risky, because "the heavy official recommendations are not profitable for the farmer."[23] Therefore, Tenenbaum suggested that farmers were applying the most economic rate of fertilizers, even if this was lower than the official recommendations.

In 1970 the deputy director of ICAR, A. B. Joshi, presented some disturbing news to the council about Kalyan Sona: "During the last *rabi* season, reports were received, regarding extensive yellowing of the dwarf wheat crops, especially Kalyansona, from U.P. [Uttar Pradesh] and parts of Bihar. The primary reason for this appears to be that farmers are not adequately fertilizing these wheats. Experimental evidence now clearly shows that the new wheat varieties, especially Kalyansona, should not be given less than 80 kg. of nitrogen per hectare and that a higher dose is definitely beneficial."[24] Despite its wide adaptation, Kalyan Sona performed best under higher fertilizer rates. Kalyan Sona also seemed to require certain conditions for optimal growth: in 1975 wheat project coordinator Rao stated that Kalyan Sona's yields seemed to be decreasing as the result of delayed sowing, lack of irrigation, and decreased soil micronutrients.[25] Tenenbaum put things more bluntly: "If 'high yielding varieties' had been developed before chemical fertilizers became available, they might long since have starved to death (whereas 'native' varieties obviously have survived)."[26]

B. S. Minhas and T. N. Srinivasan's proposal of more equally spreading fertilizers was becoming more popular among India's agricultural administrators and policymakers in the 1970s. Rao asked "whether the available fertilizer should be given to limited cropped area for intensive cultivation or whether it should be spread out both to High Yielding Varieties and tall wheats to get more yield."[27] A. B. Joshi, then director of IARI, agreed in a 1974 lecture that it was "imperative that we leave the path we followed during the late 1960s and the early 1970s" when the main goal was to increase farmer adoption of fertilizers.[28]

In 1978 engineering and development expert Kirit S. Parikh argued that "fertiliser allocation ought to be based on an analysis of local conditions and the responses of the available varieties" rather than concentrated on the high-yield-potential varieties.[29] Parikh used extensive data collected from fertilizer experiments on farmers' fields called Simple Fertiliser Trials (SFT).[30] Later, Vidya Sagar pointed out that researchers were slow to analyze results from these farm experiments, taking almost two decades to publish results.[31] The SFT results showed that farmers were not benefiting as much as expected from increased fertilizer application. Agricultural economist A. Vaidyanathan responded to Parikh's article "only to question the assertion that SFT responses are reliable approximations of the yield responses to fertiliser under conditions of widespread application by millions of farmers."[32] He explained, "The sample plots do have the benefit of expert knowledge on the mode and timing of applying fertiliser," and therefore the data were not very reliable.[33] Tenenbaum also noted the SFT plots received "'guidance' on the proper timing" of nutrients and free seed, fertilizers, and insecticides.[34] Even these farmer field trials were not representative of those farmers without "proper agronomy."

India's Seventh Five Year Plan (1985) set fertilizer targets by state, rather than only nationally, perhaps in an effort to increase fertilizer use beyond irrigated areas.[35] Even by the late 1980s, wheat scientists noted that national fertilizer use was only 25 percent of the recommended rate, and that about 70 percent of the wheat crop was receiving less than 50 percent of the recommended fertilizer rate.[36] Indian scientists attempted to prioritize "development of varieties capable of yielding well under low fertilizer and other sub normal management conditions."[37] Scientists from the Wheat Project Directorate (WPD) sent Borlaug a report to justify their use of lower fertility levels in the coordinated trials in an undated report from the 1980s:

> In earlier years trials were laid out using 120 kg.N/ha which has been reduced to 100–80 kg/ha as most of the crop in the country is sown with the use of very small doses of fertilizers. This has also affected overall performance of trials. It has been observed that the most successful varieties in India are not the ones with the highest yield potential under the best man-

aged conditions but ones which perform well over a wide range of cultural environments specifically lower fertilizer levels and fewer irrigations. . . . This has necessi[t]ated development of genotypes which can do well at moderate fertility levels.[38]

Borlaug, infuriated, responded that "none of them are justifiable reasons for low yield in experimental plots where one is trying to measure the true genetic potential of new genotypes."[39] While this was Borlaug's usual refrain, in the 1970s Borlaug had coauthored an internal report that wheat lines "must [be] tested off the experiment station, and the agronomic package must be developed under conditions facing the farmer."[40] But at seventy-two years old, Borlaug remained committed to his own vision of India's wheat development: one focused on the genetic potential of varieties under high levels of fertilizer, despite the reality on the ground.

THE LIMITS OF WHEAT RESEARCH

Anderson reviewed the Indian wheat program for the RF in 1969 and his notes were not optimistic. He wrote to Borlaug that the coordinated wheat program, which had launched informally in 1961 and formally in 1965, was not officially funded through the government until the very end of 1969.[41] The other coordinated crop programs had been funded by ICAR earlier.[42] Anderson was "literally ashamed to go out and ask the various cooperators to carry out essential actions when they have no money."[43] It is not clear why funding for the coordinated wheat program was delayed for so long.

Anderson left India in May 1971 to work with CIMMYT in Mexico, effectively ending the RF's direct support of the Indian wheat program (though it was continued somewhat through other RF and Ford Foundation field staff). On a visit back to India in November 1973, Anderson found a lack of labor for the coordinated trials and observed that plots were "only half plowed after harvest of the corn crop," not irrigated or leveled, and short on fertilizer.[44] He also found that more varieties were being released than needed due to fractured research efforts. These were symptoms of a larger problem with the new wheat research system, which Indian scientists reflected on in the late 1970s.

Borlaug had hoped that wide adaptation would hasten the speed of varietal adoption in the developing world, but Kalyan Sona and Sonalika had the opposite effect. Despite Kalyan Sona and Sonalika's quick adoption, they highlighted the gap between the varieties that scientists released and those adopted by farmers. Indian scientists released sixty-eight wheat varieties from 1967 to 1977, but only a few were sown in north India.[45] Wheat coordinator Rao wanted to know "why many of the new varieties we are recommending for release did not catch up with the farmers. Is something wrong with our present system of testing and release?"[46] He called for a "critical and unbiased review of the whole system of varietal release, multiplication, popularisation, replacement and implementation of a sound policy in this respect," though it is unclear if such a review occurred.[47]

Indian wheat scientists were aware of issues within their own wheat research system but did not consider it their responsibility to ensure the adoption of new varieties. That was considered the job of agricultural extension (this sentiment continues among plant breeders today). Yet plant breeders were not even providing necessary quantities of new varieties for the seed companies to reproduce.[48] Varietal adoption in the northeast remains slow; for example, UP262 (released in 1978) was still grown in the northeast in the 2010s. A recent analysis by Vijesh V. Krishna et al. shows that India's formal seed system still produces nontrivial amounts of older varieties that scientists no longer recommend.[49]

Poor administration and a stifling bureaucracy also limited the Indian wheat program. An internal RF report found that the Indian Agricultural Research Institute (IARI) "minimize[d] contributions of research in the coordinated cereal improvement projects made by stations other than IARI or its regional stations," in favor of research done at IARI headquarters in Delhi.[50] The same report mentioned the "duplication of effort on research both within and between Divisions. . . . Even within a given Division research on a particular crop is often far from coordinated."[51] S. P. Kohli, the wheat program coordinator, was noted to be an ineffective leader who was "being forced out by ICAR" as well as IARI in 1970.[52] Kohli soon moved to the Middle East with the United Nations Development Program and Rao, a geneticist at IARI, took over as coordinator.

The RF began phasing out the Indian Agricultural Program in the mid-1970s, believing that the Indian research system was sufficiently developed and no longer required outside intervention. This was not an adversarial breakup, and CIMMYT and the RF would continue to collaborate with Indian researchers through the international wheat program and the newly formed research center, International Crops Research Institute for the Semi-Arid Tropics (ICRISAT), in Hyderabad.[53] Ralph Cummings, the former director of the RF's Indian Agriculture Program, soon became director general of ICRISAT. The RF distributed some of their field staff from India to other RF-supported international research centers. Other staff were sent directly to Turkey, for the RF's newest national wheat program.[54] The RF program in India officially terminated in 1976.

NO GREEN REVOLUTION IN DRYLAND AND RAINFED INDIA

Meanwhile, in Bangalore in 1968 there was no celebration of the wheat revolution. Bangalore was hosting a dryland wheat conference that year. *Dryland agriculture* means unirrigated agriculture with limited rainfall. About 68 percent of India's cultivated area is dryland, and farmers here tend to be poor, with small land holdings. Kullal Chickappu Naik, the first vice-chancellor of the University of Agricultural Sciences, Bangalore, opened the meeting by remarking that "any talk of Agricultural revolution in an area subject to the worst scarcity conditions in the country and in a year when 17 out of the 19 Districts were in the grip of drought would be unrealistic, if not flippant."[55] He continued, "It is high time States like Mysore shift the focus on drought and dry farming rather than imitate and adopt programmes which are designed exclusively to help the farmers with assured irrigation facilities."[56] He declared that "Indian Agriculture can never hope to boost Indian economy through irrigated farming alone," which would lead to further economic inequity.[57] Referring to USAID administrator William Gaud's declaration of record wheat production, Naik said, "We are apt to overlook the fact that even after these bumper harvests, we have had the sad spectacle of creeping hunger gnawing at the insides of the country due to drought and floods."[58]

In other publications, scientists presented a variety of evidence showing that the widely adapted Mexican semidwarf varieties were not adapted to south and central India. Scientists from Madhya Pradesh found that the Mexican "light insensitive varieties also require a precise management, particularly timely irrigation schedule."[59] In dryland areas that dominated the state, these varieties "are not suitable. The local varieties appear to be better adapted to such variable conditions."[60] These views from scientists in dryland India were consistent with what economist Bandhudas Sen wrote on "the dependence of the high yielding varieties on irrigation."[61] He wrote: "We are not suggesting that the high yielding varieties cannot be grown at all on unirrigated land. They can perhaps be grown in areas supposedly blessed with 'assured rainfall.' But the area officially classified as under assured rainfall does not mean much . . . nor is the distribution of rainfall in these areas over the season such as to meet the exacting demands of the new varieties."[62] Sen, among others, found that the semidwarf varieties were not widely adapted across unirrigated conditions, especially in dryland India.

Initially, RF scientists had justified the lack of research for wheat varieties adapted to dryland conditions. Anderson wrote in 1966: "In the case of varieties for dry land production, less attention is being paid to

this aspect until the shortage gap is filled. It is not being completely neglected but there is simply a preference for production of varieties for
high yield under high fertility and irrigation."[63] Despite this focus on
irrigated areas, the RF scientists had some hope that the varieties developed under irrigated conditions would still be adopted in rainfed and
dryland areas. But in 1970 Guy B. Baird, director of the RF's Indian Agricultural Program, wrote, "The impact of the dwarf wheats has been
limited almost entirely to the irrigated areas; the improved technology
has not yet materially affected the two-thirds of the wheat acreage under
rain-fed (generally dry land) conditions."[64] In the late 1960s and early
1970s the RF scientists realized that their international wheat, maize,
and rice programs had reached mostly irrigated areas.

Indian wheat scientists in northern India were aware of the problems of semidwarf wheat and their limited spread to unirrigated areas.
J. S. Kanwar, a soil scientist and deputy director general of ICAR, began
corresponding with various groups in the United States that supported
dryland research around 1967. Kanwar presented his views at various
conferences, highlighting the lack of attention paid to rainfed and dryland wheat production, and the need for better irrigation systems. Kanwar found that the semidwarf varieties did not require more irrigation,
but rather more precise water management.[65] This was possible under
the tube well irrigation system prevalent in northwest India, but much of
India's "irrigated" area was not able to irrigate enough or at scheduled intervals. The wheat director's report in 1988 found that while 72 percent of
wheat was irrigated, 56 percent of this area was receiving less than half of
the recommended irrigations.[66] In 1973 Kanwar became the first deputy
director general of the RF's ICRISAT in Hyderabad, which focused on
semiarid crops, and he also helped establish the All India Coordinated
Research Project on Dryland Agriculture.

The category of rainfed agriculture is more expansive than dryland,
and about 50 percent of food crops in India are grown without irrigation. RF staff and administrators also recognized the socioeconomic inequities between irrigated and rainfed farmers, which were exacerbated
by the Green Revolution. Sterling Wortman, the RF's director for agricultural sciences, wrote to Baird in 1969: "The importance to India of
her irrigated area is quite obvious. Do you have any information on the
number of farmers who are largely dependent upon unirrigated wheat
as a means of livelihood? I think that we must increasingly consider in
our projections not only total national output, but as well, the number of
people needing benefit."[67] Cummings—at USAID at that time—wrote to
Bill Wright, an agronomist with the RF in India, to inquire about a program to evaluate "the problems of rainfed lands and small farmers."[68] RF

staff in India also mentioned that Indian policymakers were interested in trying something like Plan Puebla, a CIMMYT project for rainfed maize and small farmers, because of a "growing concern about the lot of the small farmer."[69] According to David Hopper, "knowledge of the Puebla experiment is fairly widespread among senior Government officials here."[70] The interest in rainfed farms would continue in the RF in the 1970s.

India's Fourth Five Year Plan, for 1969 to 1974, was the first national plan to mention the need for rainfed-specific research, although the plan overall reinforced concentrating resources in the northwest. This, among other factors, led to the formation of a coordinated dryland research program, proposed and initiated in 1970. At the first meeting in January 1970, Joshi emphasized "the importance of dry farming and the widening gap in production between the irrigated and unirrigated arid areas," thus "a coordinated scheme on Dry Farming has been formulated."[71] According to the diary of an RF officer, the All India Coordinated Research Program on Dryland Agriculture was launched in coordination with the government of Canada, and the initial research of the program focused on screening varieties under rainfed conditions, soil conservation practices, intercropping, and dryland crops.[72]

Later reports show the efficacy (or nonefficacy) of rainfed research in India. Wheat coordinator Rao stated at the annual wheat conference in 1978 that one of the general problems facing wheat research and production was "*raising wheat yields under unirrigated conditions. . . .* In the last several decades we must have attempted thousands of crosses to improve yield by genetic manipulations. The progress is very limited and the locals or selections from locals or varieties bred long time ago are still hard to be beaten. . . . What should be our approach to improve wheat yields under these unirrigated conditions?"[73] Rao, as we have seen through his other statements, raised some formidable questions. Also in 1978, a National Drought Screening Nursery was initiated for wheat, perhaps inspired by CIMMYT's first Drought Screening Nursery in 1976.[74] A few years later Rao again noted, "We have not done enough systematic work to produce wheats for rainfed conditions. The efforts are half-hearted, academic and, lack real thrust."[75] It was not only Rao pointing out this issue, however. The next wheat project coordinator, J. P. Tandon, noted that only a few of the "improved" varieties were adopted in rainfed areas, perhaps because these varieties "may lack some of the components of adaptability to such environments."[76]

By 1993, however, scientists were still aware that "we have given little more than lip sympathy to rainfed wheat," and asked, "Is anyone multiplying seeds of rainfed varieties?"[77] The older variety C306 was still a top-yielding variety in the coordinated trials for rainfed conditions thirty years after its release.[78] But one wheat scientist stated, "We have found it necessary to focus on the farmer in the irrigated, productive areas because they are more rewarding. We have a result-oriented program and that has to continue."[79] One could ask, results for whom? While India's agricultural researchers have given lip service to developing rainfed varieties, many scholars point out India's lack of research and investments in rainfed agriculture.[80]

The failure to help rainfed and dryland wheat farmers has many causes, many of them historically situated around Green Revolution–style research. For one, India's wheat research system is biased toward farmers in the irrigated northwest. This seems to come from the Ford Foundation's Intensive Agricultural District Programme in the 1960s that focused on irrigated and assured rainfall areas, followed by the Rockefeller Foundation's targeting of irrigated areas. The institutional structures set up around irrigated agriculture during the Green Revolution still exist now. For example, most wheat researchers and the most prestigious wheat research institutes are in the northwest. Second, Indian wheat breeders do not have incentives to produce varieties that farmers will adopt. Instead, plant breeders gain prestige from releasing varieties. Even when rainfed varieties are released, they do not necessarily enter the commercial seed system (sometimes because wheat breeders fail to submit adequate quantities of seed into the seed multiplication system). Third, researchers and administrators might argue that projects with wide national coverage should override those like rainfed and dryland agriculture, which "tend to be region specific."[81] This is an unfortunate consequence of the philosophy of wide adaptation.

CHANGING CONCEPTS IN PLANT ADAPTATION, GENETICS, AND ENVIRONMENT

India's agricultural research system is large and relatively well funded. High quality agricultural universities train Indian scientists, some of whom also receive education and training abroad. So we might ask ourselves: why was it so difficult for Indian agricultural scientists to develop wheat varieties better than Kalyan Sona and Sonalika, and C306? Why have researchers reported that the yield gains of Indian wheat varieties in the 2000s have been "very marginal" over Kalyan Sona and Sonalika?[82] Could there be some systematic flaw in the research and testing system?

I believe that part of the problem is the focus on plant breeding at the expense of agronomy physiology, and the social sciences.. Agricultural scientists have focused more on plant genetics and less on the contextual relationship between plants, the environment, and the economic circumstances of Indian farmers. And I believe the philosophy of wide adaptation has perpetuated this narrow focus. Prior to the Green Revolution, Indian wheat scientists bred varieties adapted to the prevailing agronomic conditions. In scientific terms this is now known as exploiting genotype by environment (G × E) interactions (for example, adapting the genotype for maximum yield in a specific environment). Borlaug, however, shifted the focus of Indian wheat science toward developing genotypes under favorable conditions while *claiming* that widely adapted varieties performed well in any environment. Indian scientists then began claiming that "wide adaptation is genetically controlled and that varieties do not require to be bred for the particular conditions. This indicates that minimizing G × E interaction is important for the choice of environment."[83]

Since the Green Revolution, most Indian wheat scientists have sought to minimize G × E interactions by focusing on varieties that have high phenotypic stability (low variation in yields across environments). In 1993 Tandon noted that certain research sites had high variation between years. He scolded the researchers, "When there is such a wide difference in yielding ability of genotype between any two given years, it creates more error and therefore, small genetic improvements in yield that are being made will go unidentified."[84] For this reason, Borlaug and others promoted testing varieties under the ideal environment of high fertilizer and precise irrigation. This meant that test results from rainfed or otherwise marginal environments were often thrown out.

The focus on genetics over the environment was a shift from earlier scientific thought in India. In a 1960 article, scientists S. M. Sikka and K. B. L. Jain explained that "yield in wheat, as in other crops, is the ultimate expression of the interactions between the genetic constitution of the crop variety and the environmental factors in which it is grown. The yield is, therefore, at its best when the plants of a particular variety are in harmony with the environment."[85] Further, "The production of specialized varieties to suit different agronomic conditions has recently assumed such great importance" that the Board of Agriculture emphasized this area of research.[86] Presumably, the popular rainfed variety C306, released in 1965, was developed under this research thrust. C306 was derived from the research of the celebrated Ram Dhan Singh Hooda, who had studied under Albert Howard. Hooda sought to breed varieties for the conditions faced by peasants and considered farmers as his peers. While other varieties have come to replace C306, it had a very long run and is known for its unsurpassed quality for making chapati (flatbread), something the semidwarf varieties lacked. While doing fieldwork in India in 2013, I met a farmer growing C306 for the premium organic market.

But the argument for minimizing environment and prioritizing ge-netics prevailed after the Green Revolution. Wheat-breeding research was performed under ideal conditions and often isolated from other agri-cultural disciplines such as pathology, agronomy, and physiology, much less the social sciences. Under this isolation, varieties were not bred for specific conditions. Varieties that performed the best over a range of con-ditions were promoted for release.

Kohli, as exiting coordinator of the Indian wheat program, wrote a scathing review of Indian wheat breeding in 1969 and its focus on genetics. He criticized scientists for assuming yield "to be a direct 'gene-controlled plant character,'" rather than something influenced by environment.[87] Kohli stressed the importance of the environment on yield, which he felt had been overlooked by Indian wheat breeders. He used the example of C306 versus two semidwarf varieties to show that in low-fertility environments, C306 had a higher yield than the semi-dwarfs, but that at higher fertility levels the semidwarfs had a higher yield. This indicated that there was no genetic "inherent yield," and breeders should try to improve the fertilizer response of Indian varieties and improve the efficiency of semidwarfs in low-input environments. He argued that under the prevailing mindset of Indian plant breeders, they would not be able to increase the yield of new varieties. And at least for wheat, he was right.

Kohli mentioned that the Finlay-Wilkinson model could help iden-tify useful traits in specific environments. Recall that this influential model allowed scientists to quantify adaptation and phenotypic stabil-ity (I call this and similar models "stability models"). In other words, the Finlay-Wilkinson model could quantify the G × E interactions of crop varieties. Keith Finlay himself influenced Indian wheat scientists in the latter part of the 1960s, having visited India in 1966 to lecture on adaptation and perhaps a few times after that.[88] Then in 1968 he helped organize the Third International Wheat Genetics Symposium with M. S. Swaminathan. At least two influential Indian wheat scientists, J. P. Sri-vastava, from Pantnagar, and H. K. Jain, of IARI, met with Finlay on a trip to Australia.[89]

Interestingly, there is scant evidence that the Finlay-Wilkinson mod-el was applied to wheat data in the 1960s, even though the models easily could have been applied to the coordinated wheat trial data, as I have done. It is not clear to me why. Indian crop scientists applied the model (or one like it) to millet, gram (lentil), sorghum, and rice.[90] All of these scientists worked at either IARI's division of genetics in Delhi or at Pun-jab Agricultural University in Ludhiana and Hissar, which were the ma-jor research hubs in the northwest (besides Pantnagar).[91] Then, for some reason, the 1970s brought a renaissance of studies on wheat adaptation and stability in India. Indian scientists published nearly two hundred papers on the topic for wheat alone by 1980.[92] Wheat scientists applied various stability models to intra- and interplant variation in wheat's characteristics, to see how that variation was inherited and how it was expressed in different environments.

After stability analysis entered the toolkit of Indian wheat scientists, scientists began to recognize the contradictions of wide adaptation starting in the later 1970s and into the 1980s. Kalyan Sona was considered very widely adapted due to its high yield and high rates of adoption not just across India but in other countries. When scientists analyzed Kalyan Sona and other varieties using stability models, however, "the variety 'Kalyansona' had been observed to have a high regression coefficient, indicating low stability."[93] Researchers observed, "The general experience is that this variety has wide adaptability, as it is grown throughout the world, particularly in South Asia, the Middle East and North Africa. . . . At the same time it is known to be highly responsive to high fertility conditions. . . . The wide adaptability of this variety, therefore, may not be due to its inherently greater stability but due to its distinctly greater yield potential compared with previous commercial varieties."[94] These results are not surprising, considering that Kalyan Sona was derived from wheats Borlaug had selected in high-fertility environments.

Other researchers had similar findings. Three Indian scientists presented the following at an international conference: "Kalyan Sona the best known variety from the adaptation point of view, is characterised by a relatively large G × E interaction. This finding suggests that capacity to generate variability under diverse environments may be an important attribute of a well adapted variety. The finding calls for a review of our earlier concepts on adaptation."[95] Perhaps Indian scientists would have been caught less off guard if Borlaug had shared with them not just the positive aspects of wide adaptation but also the warnings from Finlay a decade earlier. Finlay had cautioned Borlaug: "Lerma Rojo 64 is of particular interest because it is even less stable—indicating that it is better adapted to 'higher yielding' environments" and "the selection technique used at present certainly allows the selection of widely adapted genotypes but it also automatically eliminates genotypes with exceptional potential for yield given the correct specific environment."[96] Contrary to the common understanding that wide adaptation was synonymous with phenotypic stability, it seems that some widely adapted varieties had low phenotypic stability and high G × E interactions.

By the early 1970s and 1980s, Indian wheat scientists began investigating the development of varieties more specifically adapted to conditions of rainfed, drought-prone, and low-fertility agriculture. Wide adaptation had not panned out as planned to deliver varieties that yielded well in both favorable and unfavorable environments. In 1978, Rao presented findings from research on drought versus favorable environments that "the two environments bring about tremendous changes in the yield attributing characters. . . . The studies revealed that selection parameters have to be different for different environments."[97] These findings challenged the supposed inherent yield advantage of widely adapted varieties, as well as the dogma of using only one selection environment for multiple target environments. It is astounding that in 1978 it was considered a radical proposal that wheat yield was affected by environment, but this demonstrates the pervasiveness of Borlaug's argument for wide adaptation.

Still, I was shocked to discover that only in the 1990s did Indian wheat scientists start to view wheat as part of a system rather than an isolated organism. In 1993 Tandon noted that farmers were more concerned about the profits from the crop *system* rather than wheat alone. Wheat is typically grown after rice, and a late harvest of rice or delay in preparing the field will reduce the yields of wheat. Farmers in eastern India, however, consider rice more profitable and therefore prioritize it even if they get lower wheat yields.[98] Yet it took until the mid-1990s for Indian scientists to start breeding wheat for a wheat-rice rotation, such as prioritizing lower-yielding but faster-growing wheat.[99] Around that time Tandon noted that "it may be necessary to develop genotypes of different duration, nutrients and irrigation requirement," again departing from wide adaptation and recognizing the context-dependent nature of agriculture.[100] Location-specific research, an alternative to the Green Revolution style of research, has challenges as well, which we will now explore.

LOCATION-SPECIFIC RESEARCH IN INDIA

Most plant-breeding research in the developing world prior to the Green Revolution was location specific—in other words, targeted at one geographic area. Borlaug's wheat research overthrew this paradigm, but after the Green Revolution some scientists advocated a return to more localized, or at least context-sensitive, research. Advocates of location-specific research emphasize that Green Revolution research has not benefited farmers in marginal areas, who typically face high agroclimatic variability. These advocates call for more consideration of the specific agroclimatic and socioeconomic needs of farmers, especially those farmers who did not benefit from the Green Revolution.

One book from the height of the Green Revolution stood out to me for its detailed attention to the limits of the Green Revolution in less-favorable areas. Samar Ranjan Sen, a well-known agricultural economist and Indian Planning Commission member from 1963 to 1969, published *Modernising Indian Agriculture* in 1969. The report evaluated the Ford Foundation's Intensive Agricultural District Programme and brutally highlighted the failures of the Indian scientific bureaucracy to improve agriculture in India in the 1960s. Sen was adamant that research must attend to the variability of agriculture not just within states but within districts. Sen wrote extensively about the need for research "directed to local situation at the field level and in the various agro-climatic regions to meet local needs."[101] He wrote further: "To be really effective it is necessary to evolve specific programmes to suit the agro-climatic conditions in each zone and to communicate the specialized, localised aspects of this programme to all the farmers in the zone. . . . Needless to say that programmes, even if excellently prepared at the national level or state level, do not exactly fit into the local conditions at the operational level."[102] Sen and other critics had some impact on the policies in India's fourth five-year plan, of 1969. This plan included an initiative to improve the "economic conditions and correct the income disparities in rural areas through special programmes for small and marginal farmers, agricultural labour, drought prone areas and dry farming areas."[103] Over the next few decades, calls for more location-specific research entered the discourse of Indian agricultural science policy, especially from economists.

Post–Green Revolution critics also brought attention to location-specific agricultural needs and how a centralized bureaucracy did not serve these areas well. In the mid-1970s economists K. Kanungo and P. E. Naylor wrote, for the World Bank, a scathing review of the coordinated crop improvement programs. They found that the "common denominator of the gaps identified in almost all types of research activity is the weakness of the link between research and the actual needs of farmers in their specific location and environment."[104] According to Kanungo and Naylor, the new coordinated system had in fact centralized research and usurped local research efforts.

Agricultural scientists had little incentive to address local issues, despite agriculture long being considered a "state subject" in India. Kanungo and Naylor proposed developing decentralized, regional research facilities "on the basis of agro-climatic suitability in order to strengthen a reorganized" research system.[105] Based on this review, in 1979 the World Bank and ICAR started the National Agricultural Research Project (NARP), which continued until 1996 after being renewed once in 1986.[106]

NARP was created in response to the limitations of Green Revolution–style research, which was highly centralized. NARP aimed to support regional, location-specific research through the state agricultural universities and extension.[107] It promoted interdisciplinary research and aimed "to give special emphasis to cereals, pulses and oil seeds under rainfed and mixed farming conditions."[108] The summary of NARP II (after its renewal in 1986) pointed out that "there is scope for greater progress in adapting the new technology to suit the varying socio-economic conditions of farmers and in meeting the specific needs of the diverse agroecological zones encountered in a country as large as India."[109] One of NARP's first products was a map of 127 agroclimatic zones within India. A research station was established to serve each zone and undertake "adaptive" research.

In a similar vein to NARP, in 1988 India's Planning Commission created the Agro-climatic Regional Planning project to support location-specific research and development as part of the Seventh Five-Year Plan. This program was based on the relatively new "farming systems" approach and emphasized long-term agricultural sustainability.[110] The project delineated fifteen agroclimatic zones and created Zonal Planning Teams to provide input to the Planning Commission. The Zonal Planning Teams appear to still operate, and in 2021 Prime Minister Narendra Modi "urged states to prepare agro climatic regional planning strategies to help farmers, and create conditions for boosting exports at the district level," especially for oilseeds.[111]

Unfortunately, NARP had little impact due to lack of sustained investment. For example, there is little to no discussion of NARP in the annual reports of India's wheat program. Perhaps there are documents or text that I missed, but one reference I found from 1993 stated, "During the year a new project on strengthening of research on rice-wheat system with the world bank aid (NARP II) was initiated at ten centres."[112] There is no discussion of how climate-specific research would be integrated into the wheat research program in the public reports that I studied. With some exceptions, agricultural research in India remains extremely centralized and disciplinarily isolated.

A report by Panjab Singh for the Agro-climatic Regional Planning project in 2006 explained some of the failings of NARP. First, the more than three hundred Zonal Research Stations created for adaptive research "in many locations are non-functional institutions in absence of adequate funds after the withdrawal of the World Bank support."[113] Second, there were two issues with the research orientation of NARP. First, NARP focused more on crops than on farming as a system, and "in the process rainfed agricultural research suffered and did not get the emphasis it deserved."[114] NARP also focused more on technology *generation* than on *transfer*, which led to the generation of several technologies that were inappropriate for farmers' contexts. Finally, "the research work done in NARP generally did not, by and large, consider" issues that are salient for the poor, women, or the environment.[115] These results suggest that despite the intentions of NARP to address location-specific research, it did not address the institutional barriers to this type of research.

Discussions over wheat research went back and forth on the continuum from location-specific to widely adapted research in the 1980s and 1990s. This was consistent with broader trends in the agricultural research community in the 1980s, which attempted to address the problems of marginal farmers through locally adapted research.[116] In a book celebrating the twenty-fifth year of India's coordinated wheat program in 1986, then director Tandon and former coordinator Rao wrote, "[Until] recently the workshops considered proposals for the zonal releases only. But now it has been decided to identify varieties with more specific adaptability also for a state or parts of it."[117] On the next page, however, they revealed their still-skeptical stance toward decentralizing research among India's states, stating:

> Although very elaborate and extensive system of thorough testing and identification of suitable varieties has been developed under the AICWIP [All India Coordinated Wheat Improvement Program], yet some of the states continue to organise regional based programmes. The utility of such programmes has been questioned but agriculture being a state subject, it can not be banned. Moreover, possibilities of developing varieties with very specific adaptability cannot be denied and some times there may be a need to develop varieties for specified regional problems. However, it is interesting to note that examples of state released varieties having become popular any where are very rare.[118]

Although Tandon and Rao seem to have appreciated location-specific research to address "specified regional problems," they did not view it as particularly desirable. They also dismissively referred to the state mandates for agriculture by saying simply that "it can not be banned," implying that agricultural research was better coordinated on a national or zonal level.

A few years earlier had Tandon written, "It was felt that the extent of germ plasm entering wider adaptability test is very restricted and needs broadening. Keeping this in view, a National Yield Screening Nursery was organized" for the wheat program.[119] In the twenty-fifth anniversary book, R. K. Agrawal also lamented the lack of new widely adapted varieties. He observed that it seemed either the Indian wheat breeding programme had "not succeeded to develop varieties of wider adaptability like Kalyan Sona and Sonalika for irrigated condition and like C306 for rainfed cultivation" or these varieties had not been entered into national trials.[120] Over the 1970s and 1980s the Indian testing system attempted several times to hold national trials to identify widely adapted varieties, but lacking success, this effort was ended and transformed into a National Yield Nursery.

Still, well into the 1980s and 1990s Borlaug was certain that trials should be conducted at high fertility and aimed at wide adaptation. Borlaug wrote to Tandon in 1984: "The fact that some varieties such as Sonalika, Kalyan Sona, HD2009 and WL711 are grown commercially over wide areas, indicates that it is possible to breed varieties with broad adaptation. . . . This would seem to me to indicate that there are altogether too many zones that are used in the varietal release programs and that these could be greatly simplified by fusing a number of smaller ecological zones."[121] By 1984 Indian wheat scientists had expanded the initial five agroclimatic zones into nine to accommodate hilly areas and other specific environments. Thus, while Indian scientists recognized the limits of wide adaptation, Borlaug continued to push wide adaptation and the consolidation of agroclimatic zones.

Meanwhile, the rice research system in India became much more decentralized and location specific. After the success of wheat in Mexico, the Ford and Rockefeller Foundations founded the International Rice Research Institute (IRRI) in the Philippines. This institute and its research model were based on CIMMYT and aimed to develop short-strawed rice varieties that produced high yields under irrigation and fertilizer.[122] The first variety they released, IR-8, was claimed to be a widely adapted variety. IRRI hailed it as "a rugged variety that could go almost anywhere."[123] IR-8 was released in India in 1967, but "some national scientists, India's in particular, questioned IR8's adaptability. Indian scientists felt that a variety should be tested widely before being released to areas outside of where it was bred and raised."[124] This is consistent with India's wheat scientists' views on widely adapted wheat.

But IR-8 and other IRRI varieties did not have the success of semidwarf wheat in India; they languished under environmental stress, pests, and disease.[125] Both Indian and IRRI scientists "realized that 'wide adaptation' wouldn't work in rice, and that varieties needed to be more tailored to local agroclimatic conditions."[126] Borlaug's model did not translate to rice because rice is grown in much more diverse conditions than wheat is. Edmund Oasa found that for IRRI scientists, wide adaptation "actually meant wide adaptability under controlled conditions" but that scientists recognized they needed a new strategy.[127] In the 1970s and 1980s IRRI and Indian scientists reoriented rice research toward local adaptation.[128] By 1982 the All India Coordinated Rice Improvement Program had "recognized in recent years that paddy breeding research must be decentralized to become location-specific."[129]

A similar reorientation did not occur in wheat research. Indian wheat research has struggled to conduct location-specific research, partly because of the centralization of Indian agricultural research in general. Even ten years after NARP was implemented, agricultural research was still very centralized. A 1989 report on the state of India's state agricultural universities found that the "government requires all donor funds to be channeled through ICAR, which delays or even prevents extramural funding of projects. In addition, it means that research projects that receive high priority are the ones which have a broad national coverage."[130] The authors highlighted "the need to decentralize decisions concerning research budgets," and concluded that the existing system was biased against rainfed agriculture.[131] Even with repeated calls to focus research on rainfed and dryland areas, these efforts have had only marginal success.

Science policy scholar Rajeswari Raina has written extensively on the barriers to institutional change in Indian agricultural science. She argued, "By the mid-1970s, agricultural science had become one of the administrative inputs locked into India's food security goals by weeding out the variety or organizational formats and sources of funding and eulogizing the achievements of the green revolution" to the detriment of state- and district-level agricultural development.[132] Raina has written that repeated calls for change have failed because they do not address the existing norms, values, and incentives of the research system. She has suggested that scientists and administrators need to adopt new norms that encourage the researcher to "cater to the local farming communities, respond to different biotic and abiotic stress, and foresee and warn the state and stakeholders . . . about possible risks."[133] Raina's research helps explain why the attempts to decentralize Indian wheat research have failed: they have not addressed the core norms and values in Indian wheat science, which continue to incentivize centralized, isolated research.

NEW REVELATIONS FAIL TO REFORM THE GREEN REVOLUTION MODEL

In the decade that followed the RF's intervention in Indian wheat research, the initial excitement over wide adaptation faded and troubling questions emerged. The dominance of the varieties Kalyan Sona and Sonalika brought up doubts about the Indian wheat research system and questions about why new Indian varieties were not catching on among farmers. Indian scientists and economists publicly grappled with earlier questions around fertilizer distribution and whether the concentration strategy was justified. Scientists publicly questioned the suitability of the semidwarf varieties in dryland and rainfed areas. These scientists have called for more attention to rainfed research since the 1960s, yet there seems to be little progress.

In the 1970s Indian scientists revisited the very idea of wide adaptation after discovering that Kalyan Sona had lower phenotypic stability than other Indian wheat varieties. This presented an apparent paradox because farmers had quickly adopted Kalyan Sona and it remained a popular variety for many years. Rice research in India followed a different trajectory after scientists realized that rice was not as adaptable as wheat, in part because of the genetic differences of rice and in part because rice-growing environments are more diverse than wheat-growing areas.

Nonetheless, wide adaptation was codified in the Indian wheat research and testing system. Even as national and international research goals have shifted toward promoting equity of rural populations and alleviating poverty, the Indian wheat research system has remained focused on irrigated, favorable conditions. This system has resisted attempts at reform and continued to pursue wheat research focused on genetic improvements, especially for yield. Yet few wheat varieties developed by Indian scientists have had the success of the initial Green Revolution varieties. Further, expanded wheat production came at the expense of pulses, oilseeds, and coarse grains. Unfortunately, India's agricultural science policy continues to reinforce the Green Revolution model of wheat research.

CHAPTER 4

THE PERSISTENCE OF WIDE ADAPTATION IN INDIA

Having thus far examined the historical origins of wide adaptation and its role in shaping agricultural development before and after the Green Revolution, I now turn to the present context of agricultural research in India, the role of the wide adaptation concept in agricultural research policy, and critiques of the Indian agricultural research system. There is a mismatch between the agricultural research goals and the food security goals in India. In other words, the supply of research and new innovations does not match the demand for more equitable access to nutritious food. I argue that wide adaptation is still fulfilling the role it did in the Green Revolution. Wide adaptation is used to justify a focus on favorable farming conditions by supporters who claim that its benefits will trickle down to more marginal areas and that net gains in production will inevitably lead to a food-secure nation.

In early 2013 I attended a science policy workshop in New Delhi attended by Indian science and technology scholars and activists. A refrain established early in the day was, "there is no Indian science policy," meaning there is no coherent vision to guide Indian science toward the goal of improving society. Barry Bozeman and Daniel Sarewitz have defined science policy as political "efforts to bring technical knowledge to fruition and application."[1] The same can be said for Indian agricultural science policy: while certain policies and budget allocations affect agricultural research, there appears to be little overall strategy connecting agricultural research with specific desired outcomes. India's agricultural policies were until recently guided by its Five-Year Plans, which were developed by the Indian Planning Commission. The Indian Council of Agricultural Research (ICAR) and the Department of Agricultural Research and Education (DARE) then execute those plans.

Recent agricultural goals laid out by the Planning Commission include increasing the annual growth rate of food production and giving increased attention to rainfed agriculture.[2] The agricultural growth goal is to produce more food, but it is also based on the assumption that "agricultural growth is a means to the larger goals of employment-led growth and poverty reduction."[3] India's Eleventh Five Year Plan (2007–2012) recognized that the Green Revolution had not benefited all farmers equally and appealed for a more demand-driven agricultural research agenda.[4] This coincided with the "Bringing the Green Revolution in Eastern India" campaign to improve crop yields in eastern India, which started in 2010–2011. The Eleventh Plan also attempted to decentralize agricultural funding to encourage state and regional initiatives, to aid smallholder farmers through land reform and access to credit, and to improve the focus on food security. Five years later, however, the Planning Commission recognized that "there appeared to be lack of any clear agricultural research strategy or [plan] to assign definite responsibilities and prioritise the research agenda rationally."[5] Despite the stated goals of the Eleventh and the subsequent Twelfth Five Year Plan, it was evident from my interviews with Indian researchers and administrators that India's agricultural goals are mainly perceived as improving national production and yields, with scant attention to equity and food security.[6]

The "Bringing the Green Revolution in Eastern India" campaign recognized that despite the spread of Green Revolution varieties to eastern India, yields there are much lower than in the northwest breadbasket. Poverty and food insecurity are also persistent problems in this region. Eastern India, which covers seven states, is characterized by rainfed agriculture and small farms. Rice and wheat are commonly rotated, especially in the Indo-Gangetic Plains region, but this diverse region also grows a variety of other grains, pulses, oilseeds, vegetables, fruits, and spices. While some irrigation is available, it is typically not adequate (in timing or quantity) for wheat's irrigation requirements. Yields in the northeast are also hampered by lack of nutrient management, terminal heat stress, and slow varietal turnover owing in part to a lack of adapted varieties and poor dissemination of adapted varieties.[7] The campaign appears to have had some success in raising rice yields in eastern India primarily through technologies such as irrigation and mechanization and improving farmers' access to markets.[8]

Recent research, however, has decoupled agricultural production and poverty alleviation. Santonu Basu wrote that promoters of the Green Revolution assumed that "the so-called 'trickle down' effect would take care of the poverty" but that only the breadbasket states benefited as a result of assured irrigation.[9] Many experts had assumed that the Green Revolution would increase the demand for labor and therefore raise wages. The evidence of this impact is mixed. Donald Freebairn, an economist who worked for the Rockefeller Foundation's Mexican Agricultural Program (MAP) in the 1950s, found that even decades later, in 1995, there was no academic consensus on whether the Green Revolution increased or decreased inequality, though most studies with a conclusion found that inequality had increased.[10] Even the World Bank's Operations Evaluation Department found that "cereal-based farming alone does not seem to be a solution, at least in unirrigated areas, to create more jobs or, for that matter, to alleviate poverty and hunger, which remain a typical phenomenon of rural India."[11]

In a more modern context, David Harris and Alastair Orr conducted a literature review of crop improvement programs in rainfed areas and found that while these programs can improve household food security, they have marginal income gains. Studies in semiarid India found that "while between 1975 and 2004 average income per capita rose by 114%, only 4% of this increase came from agriculture and only 1% came from crop production."[12] The authors concluded that "the primary impact of the Green Revolution on poverty in Asia was to reduce the share of household income spent on food by effectively halving rice prices," which is certainly significant but does not make agricultural production a "pathway from poverty" for rainfed farmers.[13] To echo Freebairn's remarks from his 1995 article, this is not to argue whether increased production is good, but whether it is leading to the desired outcomes.

Indian agricultural policy still relies on the Green Revolution model despite widespread acknowledgment that this has led to inequities. This model is not capable of promoting environmental sustainability or nutrition security in the twenty-first century. Rajeswari Raina, perhaps the most incisive scholar of agricultural science policy in India, wrote: "Food security in the country now depends crucially on giving up the '1970s model of agricultural development,' based exclusively on cereal production, agricultural research to test and release varieties (especially of wheat and rice) and subsidies and price supports that encourage mindless exploitation and degradation of natural resources."[14] She presented an alternative model of "integrated food, agriculture and nutrition policy" that was articulated by economist Vijayendra Kasturi Ranga Varadaraja Rao in the 1980s.[15] Rao promoted the goal of "the minimization of nutritional inequalities among the people" rather than agricultural growth.[16] It is along this framework that I evaluate the failures of agricultural science policy in India.

The effective goals of Indian agricultural research are to produce new varieties, technologies, and techniques that farmers will apply to increase aggregate production. To that end, the research system appears to meet its goals: the All India Coordinated Research Project on Wheat and Barley has released around 450 varieties from the Green Revolution onward, and wheat production has increased 4.72 percent per year since 1950.[17] Suffice it to say that new varieties and higher production do not lead to food and nutrition security. According to Raina, "policy obsession with food production targets continues[,] when there is scientific evidence that even with record food production, food security and nutrition security are getting worse."[18]

While the Green Revolution did increase staple grain production and has kept food prices low in Asia, we cannot ignore the point that currently, "around two-thirds of the world's undernourished live in Asia, the continent where the Green Revolution claims its greatest success in terms of yields."[19] Either the Green Revolution was not as successful as the narrative claims, or Green Revolution–style agriculture has failed to address the modern challenges to food security. India itself holds one-quarter of the world's hungry population.[20] Despite the scientific advances in agriculture in the past century and adoption of irrigation, fertilizers, and new varieties in India, the country still has high rates of child malnutrition. India struggles with malnutrition even as poverty decreases.[21]

There are many reasons for the contradiction of ever-rising grain production with food insecurity. One reason is that the Green Revolution increased staple crop production at the expense of indigenous and more nutritious crops such as coarse grains and pulses. Crops like oilseeds, millets, and pulses are grown in rainfed areas, which are underserved by the research system.[22] Coarse grains like millets and sorghum are not as profitable as wheat and rice but are more nutrient dense. Similarly, mixed-cropping systems can be more nutritious and climate resilient but are often considered "backward" by researchers.[23] These factors have contributed to the triple burden of nutrient deficiency, hunger, and noncommunicable diet-related diseases such as obesity.[24] Further, lack of access to food is mediated by social and political factors, which is why famines can occur even in times of food surplus.[25] The roadblocks to food security in India are complex and deeply embedded. These roadblocks occur at nearly every step of the food system, despite government policies meant to encourage cereal crop production and subsidize food-insecure populations.

Another complication in India's food security landscape are the wheat production and food security policies, which contribute to inequalities between regions. Most wheat is produced in the northwestern states Punjab and Haryana, which have higher socioeconomic status. But that wheat is not fully accessible to rural consumers, due to a lack of facilities to store and transport wheat (leading to massive stockpiles) and underinvestment in market infrastructure.[26] In the northeast state of Bihar, for example, only 18 percent of people received a full entitlement of grains through the public distribution system shortly before India's National Food Security Act of 2013, although states have implemented reforms to improve distribution.[27] Of course, access to food is not the only cause of hunger and malnutrition: lack of sanitation, childhood disease, gender inequity, and diet are important factors.

Should we hold agricultural scientists accountable for this disconnect between the nutritional needs of the country and the research they pursue on high-yielding staple crops? The research community itself can claim to be only following the research policies set by politicians. But by understanding the values, incentives, and bureaucracy of the Indian agricultural research system we can see how the Green Revolution narrative fails to address food and nutrition security.

Wide adaptation is the crutch that holds up the narrative of the Green Revolution: if scientists can claim that crops are widely adapted, they can also claim them as scale neutral. Wide adaptation and scale neutrality mean that scientists can ignore the biophysical and socioeconomic variation present in agricultural systems and design technologies for the "proper" farmer. This is a value-based assumption that is the result of decades of momentum behind the Green Revolution and a failure of policy to meaningfully address its shortcomings. In this chapter I examine why Indian wheat research is still based on wide adaptation despite the recognized limits of this strategy.

CONSEQUENCES OF WIDE ADAPTATION IN INDIAN AGRICULTURE

Scholars have argued that the dogma of wide adaptation and the strict bureaucratic structure of Indian agricultural research have stunted the system's capacity to innovate for a diverse set of farmers, crops, and environments.[28] I agree with this analysis and argue that the Indian wheat research system is so biased toward ideal conditions that it often fails to produce useful innovations. This leaves Indian agriculture vulnerable to both climate change and environmental degradation.

Wheat research in India is overwhelmingly conducted in the public sector. The All India Coordinated Research Project on Wheat and Barley (AICRP; formerly the All India Coordinated Wheat and Barley Improvement Program, AICW&BIP) is the coordinating body for wheat research in India. Under this project, India is divided into six wheat-growing zones. These zones demarcate regions for the testing, classification, and release of wheat varieties. Despite the zonal orientation of India's coordinated wheat research, agriculture is also considered a state subject. Thus, each state has a state agricultural university. State agricultural universities, especially those in the northwest, are major centers of wheat-breeding activities. State breeding activities are typically aimed at the state level, while ICAR-funded and Indian Agricultural Research Institute (IARI)-affiliated institutes work on the zonal level.

The ICAR–Indian Institute of Wheat and Barley Research, formerly the Directorate of Wheat Research (DWR), is India's national wheat research center, located in Haryana. I will refer to the DWR and AICW&BIP throughout this book because those were the names used during the period I studied. The DWR focused on research such as crop improvement (plant breeding and biotechnology), crop protection, crop quality, resource management, and social science. The DWR also coordinated the system of wheat varietal testing and release. New varieties developed by the DWR, state universities, or other organizations are tested under multiple locations and conditions for several years. Varieties are approved for release based on agroclimatic zones or at the state level. Once varieties are approved, the institute of origin is responsible for replicating the variety into "breeder seed," which then goes through several more replications by state organizations and the National Seed Corporation, based on estimates of seed demand. Seeds are then distributed and sold through these and other organizations. Because wheat is a self-pollinating crop and seeds can be saved from year to year, only about 20 percent of farmers purchase new seeds each year, and fewer than that switch to new varieties.[29]

The success of the Green Revolution depended on a massive deployment of a public extension program. Today, only the more affluent states have functioning agricultural extension programs, and resource-poor farmers typically have less access to extension. Extension's recommendations are also typically less relevant to the circumstances of marginal and poor farmers. The public-sector system of research, seed production, and marketing operates effectively in Punjab, Haryana, and western Uttar Pradesh, but the situation in northeast and central India is quite different.[30] In the northeast, extension efforts are concentrated around the research stations, and the main source of information for most farmers is from seed and fertilizer sellers. Suresh Chandra Babu et al. reported that in 2003, "sixty per cent of the farmer-households in India did not access any information on modern technologies that year."[31] But even if India suddenly invested in a more robust research-to-extension pipeline, this would not solve some of the problems fundamentally constraining India's agricultural innovation system.

A main goal of the AICW&BIP was to release new higher-yielding varieties. But despite the 450 wheat varieties released since the Green Revolution, only a handful gained popularity and dominated for stretches of several years. This leaves Indian wheat production vulnerable to disease and abiotic stresses (such as drought and heat). Many of these popular varieties are derived from CIMMYT's wheat materials, which leads one to further question the efficacy of the Indian system.[32]

Researchers V. V. Krishna, D. J. Spielman, and P. C. Veettil found that farmers were adopting new varieties *less* frequently in 2009 than in 1997.[33] One reason for this slowdown is that new varieties over the past few decades have not had drastic yield improvements; thus, farmers have less incentive to purchase new seeds and they continue saving older seeds. Varieties are replaced even less often in rainfed and marginal environments, where farmers tend to be poorer and face more variable climate conditions. In these environments, farmers lack access to quality seeds and varieties that are suitable to their growing conditions.

Marginal and rainfed farmers lack well-adapted varieties for two major reasons: (1) the strongest wheat research programs are in the northwest and focused on ideal conditions, and (2) seed systems fail to produce adequate quantities of recent, well-adapted varieties.[34] Krishna, Spielman, and Veettil showed that even in northwest India, only a few older seed varieties are multiplied and distributed.[35] The lack of new varieties also points out the poor linkages between research, extension, and seed production. Krishna, Spielman, Veettil, and S. Ghimire wrote: "First, the research system may not be efficiently identifying and translating farmer preferences for varietal attributes into cultivars that they are willing and able to adopt. Second, the seed production system may not be producing what farmers actually demand due to a poor seed demand assessment system. Third, extension programs and other distribution mechanisms may be underperforming in their efforts to convey the genetic superiority of improved varieties to farmers."[36] In my research I observed that farmer preferences in varietal adoption were seldom discussed. There was a near-universal agreement that the strongest farmer preference was for yield—thus, the entire research system is built around evaluating varieties for their yield. Yet studies have shown that farmers do not only value yield, and that they will accept lower yields in exchange for higher yield stability over time.[37] These nuances deserve more attention in India and likely require research methods more complex than asking farmers their preferred varietal traits.

The AICW&BIP used a strict method of multiyear, multilocation trials to test and approve new wheat varieties. This system was founded on the idea that varieties should perform well over a variety of environments, literally the tenet of wide adaptation. Unfortunately, this means that varieties adapted to a specific or local environment are weeded out and not recommend for release. For a variety to be approved by the central varietal release committee, it must be tested for adaptation across an agroclimatic zone, which spans several states. While varieties with more specific adaptation can still be released at the state level, the central system must still "notify" the seed for quality, which requires additional years of testing in the multilocation trials.[38] The 1998 book *Seeds of Choice* examines the Indian varietal testing and release system in detail and critiques it for not providing farmers with a choice of varieties suited to their needs. While this book is now over twenty years old, most of the critiques still apply.

The multilocation testing system is intended to ensure that only high-quality varieties are released in India, but the process can automatically eliminate varieties that are suited to specific conditions. Therefore, the varieties that are released tend to perform best in ideal conditions. The multilocation trials test varieties under different conditions such as irrigated versus rainfed, timely versus late sown, and so on. But there are still major differences between trial conditions and those of farmers' fields, especially small and marginal farmers who constitute most of India's farmers—about 85 percent.[39] Northeastern India requires varieties with drought resistance, heat tolerance, and a shorter time to reach maturity. But these varieties are not available in part because they are screened out in the multilocation trials for low yields.[40]

The multilocation trial concept is an example of how scientists attempt to minimize environmental variation through carefully controlled trials. Rebecca Nelson, Richard Coe, and Bettina Haussmann wrote that agricultural researchers usually have "an implicit assumption that farm conditions are uniform and favourable. Research is conducted on research stations under conditions that do not resemble those of smallholder farmers, then perhaps repeated on a few farms. The results of a few trials are averaged, and recommendations are made on that basis."[41] This is not unique to India and is perpetuated in many international and national contexts.

Based on the Green Revolution origins of the multilocation trial system, it is not surprising that experts feel that "the present public seed system [is] designed to promote high input responsive varieties bred by the research institutions."[42] There is also new evidence that plant breeding systems are biased not just to ideal environments but to agroclimatic characteristics of the research site. A study of agricultural household data in Nigeria found that "agricultural productivity and technical efficiency" was "significantly positively affected by the similarity of agroclimatic conditions between locations where agricultural households are located, and locations where major plant breeding institutes are located."[43] This was true even when socioeconomic factors and distance to the research stations were controlled for. In other words, farmers growing under conditions more similar to the research stations derived more benefit from agricultural innovations.

Participants of a recent workshop in India criticized the current breeding system and called for "an appropriate seed system designed to meet the diverse requirements of rainfed areas such as local adaption, climate resilience/risk minimization," and flexible seed systems.[44] These needs are not met by the multilocation trials and broader research system. The authors of *Seeds of Choice* proposed that varietal tests should occur in realistic farming environments, ideally farmers' fields, with a greater diversity of environments represented. They also proposed that farmers could provide feedback on post-harvest traits such as taste, quality, storage, and fodder use. A more flexible breeding and testing system could produce varieties better suited to farmers' needs.

During the Green Revolution and its aftermath, many political discussions focused on levels of fertilizer that were most appropriate for research and testing. As fertilizers have become more available, the social justice focus has shifted to rainfed agriculture. Rainfed agriculture accounts for 50 to 60 percent of India's agricultural area and contributes to about 40 percent of production, yet it has been historically overlooked by research programs. India has the most rainfed agriculture in the world but some of the lowest yields in these areas. Most rainfed farmers are also small farmers with less than two hectares of land, and rainfed regions are more vulnerable to the effects of climate change. Some researchers and advocates call for more investment in India's rainfed areas, especially the northeastern region that suffers from several problems, including low crop productivity, lack of electrification, and pervasive poverty.[45]

This movement differs from the "Bringing the Green Revolution in Eastern India" campaign, which aims to replicate the conditions of the initial Green Revolution. These advocates, in contrast, call for more location-specific research for rainfed environments. According to a coalition called the Revitalizing Rainfed Agriculture Network, "state-directed policies relating to the Green Revolution have resulted in a situation in which any kind of state support for the agriculture sector becomes effective only when there is availability of water for agriculture."[46] Authors from the network argue that the Green Revolution policies and technologies cannot be simply transferred to rainfed areas; new institutional systems are needed. Even agricultural researchers acknowledge that "there is no targeted programme to tackle wheat improvement work for rainfed areas" and that these areas still rely on technological spillovers from irrigated agriculture.[47] Despite these calls for change, the Indian wheat research system is still very biased toward irrigated areas. An article from India's *Economic Times* reported that ICAR allocated only 13 percent of its budget to rainfed research and that rainfed areas receive only 6 to 8 percent of national agricultural subsidies.[48] These figures suggest a severe underinvestment.

Most of India's wheat production is classified as irrigated, but this statistic can be misleading. Only about one-third of Indian wheat production can be considered "fully irrigated," meaning the recommended number of irrigations are applied at the recommended times.[49] In the northeast Indo-Gangetic plains, regions classified as irrigated may be able to apply only one to two irrigations per wheat season, which is much lower than the recommended four to six. A prominent research administrator in India recalled being informed by many Indore farmers that they needed a wheat variety suited to only one or two irrigations per season. He stated, "In their opinion, tests conducted with 5 to 6 irrigation is a luxury which they cannot afford."[50] While some researchers have debated the efficiency of research investments in rainfed and marginal lands, statements from both scientists and advocates show that these areas cannot continue to rely on technological spillovers from irrigated agriculture.

In the past, scientists could justify their focus on irrigated agriculture as part of the Green Revolution's "betting on the strong" approach. They used the concept of wide adaptation to rationalize that rainfed and marginal farmers would also benefit from the trickle-down effects of genetic improvement of varieties. Fortunately, some improved varieties of wheat also benefited nonirrigated areas. But today, India's main wheat-growing region faces groundwater sustainability issues. The Indo-Gangetic Basin's groundwater level has dropped significantly as a result of agricultural use, which leads to salinity in the west and arsenic contamination in the east.[51] Irrigation expansion may be limited due to these issues. Climate change is also a threat to wheat, which is already hampered by heat stress in the northeast.[52] Too much of an increase in temperature near the harvest time leads to shriveled wheat grains, which happened in northwest India in spring 2022 due to an intense heat wave. This is called terminal heat stress.

Climate change highlights the shortcomings of the current Indian wheat research system. South Asian agriculture is predicted to be more vulnerable to heat and water stress from climate change than any other region.[53] Up to 50 percent of areas currently classified as "favorable" environments for wheat in South Asia will become heat-stressed.[54] While wide adaptation theoretically would be helpful in buffering the impacts of climate change, the current system produces varieties for favorable conditions and systematically excludes varieties with specific stress-tolerant characteristics, because these generally have lower yields. India's mandates for research and testing provide limited avenues for scientists to develop stress-tolerant varieties. This is not a sustainable research system despite the bureaucracy it has built over the past half century.

VIEWS FROM THE FIELD

While studying government proclamations is useful to show the evolution of agricultural policies in India, I believe there is great value in analyzing the perspective of agricultural scientists themselves. These scientists are, after all, beneficiaries of the consolidation of power that occurred during the Green Revolution.[55] The Indian agricultural research system places scientists, especially plant breeders, as its most important actors and they have a great deal of power in shaping the direction of agricultural science policy in India.

In 2013 I conducted nearly fifty interviews with current and retired wheat scientists from a variety of research institutions in north India. All of my interviews took place in the Northwestern and Northeastern Plain Zones, which are the primary wheat-producing regions of India. The Northwestern Plain Zone includes the states of Punjab, Haryana, and Delhi, parts of Rajasthan and Uttar Pradesh, and the nonhilly areas of the northern states. The Northeastern Plain Zone includes the eastern part of Uttar Pradesh, Bihar, Jharkhand, West Bengal, Orissa, Assam, Sikkim, and the nonhilly areas of the eastern states.

I interviewed Indian wheat scientists spanning the major agricultural research fields: plant breeding, biotechnology, genetics, plant protection (pathology), quality, agronomy, and extension science. These interviews covered a broad range of topics related to wheat research in India (see the appendix for more information). The influence of the RF and the Green Revolution narrative on Indian wheat research is clear. Many of the scientists' responses reflected themes related to wide adaptation. Wide adaptation very obviously remains a major goal of the wheat-breeding program, both ideologically and through the policies and organization of the AICW&BIP. An interview respondent stated a common view: "In the All India Coordinated program our aim is to have varieties with the wider adaptability." Other scientists reflected that "every breeder likes to go for developing the material which can fit everywhere." Several scientists supported research on a wide adaptation basis only; for example, "In my opinion we should breed for larger area. . . . The conditions for wheat cultivation do not vary much," and "We need varieties that should not be very specific to any particular location but should have a wide adaptability." Others noted that even though their research focused on the zonal region, "it applies to whole India."

Breeding methods and philosophies also reflected the historical influence of Norman Borlaug's view that a widely adapted variety would be more stable under different conditions. Under this perspective, scientists view genetics as the main contributor of yield stability across conditions. While I did not ask my respondents to define wide adaptation, they generally referred to a widely adapted variety as one that can be grown over a wide area and that is resistant to different types of stress. For example, "If a variety's having wider adaptability, it means it is not much influenced, say, by climate." One scientist stated, "Whether it is due to disease, biotic stresses, abiotic stresses, whatever the stress is, we breed so the losses to the stresses are minimized because of inherent capacity of the variety." Another scientist mentioned shuttle breeding to produce climate-resilient varieties, stating, "Our main challenge is to select and breed the varieties which are adapted to wide range of climates. So being climate-resilient . . . a particular variety can yield higher under higher temperature regimes, under low irrigation, or even after if the cooler period is gone. We are using this shuttle breeding program to select such varieties." These responses show a strong adherence to the philosophy of breeding for wide adaptation.

The interviews revealed that most of the wheat breeders were "working for the wheat improvement for optimal environment." One scientist noted that "better-performing genotypes having wider adaptability are selected on the mean basis only. Environment effect is not taken into account." The same scientist stated that "varieties developed for high-yielding environments" did not yield well for marginal farmers only because "they're not putting input [fertilizer, irrigation, pesticides, and the like] so they're not getting yield." The statements quoted here are extremely resonant of the Green Revolution–era arguments seen in chapters 1 and 2. They also reflect a bias toward the genetic capability of varieties and a neglect of the environmental context of yield.

To understand what wide adaptation meant to Indian wheat scientists, I asked each scientist about the targeted scale of their wheat research; for example, national, zonal, state, or district level. Responses could be coded from twenty-five of the interviews I conducted. This

TABLE 4.1. Coded responses to interviews with Indian wheat scientists for the question: Does your research focus on a wide area or specific location?

Research focus	No. of responses
Specific conditions	2
State	3
Agroclimatic zone	15
General conditions	5

question and a follow-up question about breeding for specific climates/ environments elicited nuanced and extremely varied responses from the scientists on the present state of location- or condition-specific research as well as their personal views on the topic. I let scientists define terms for themselves, because the meanings of terms, even of *wide adaptation*, are not static. To some scientists, breeding for a specific location meant the district or even subdistrict level. To others, a specific location could mean a state or multistate agroclimatic zone.

As table 4.1 shows, most of the wheat scientists that I interviewed worked on a zonal basis. This included a cross-section of scientists at different institutional levels, including state universities. The five scientists who were working for general conditions—that is, the whole country—were at the DWR. One scientist there said, "Our mandate is complete India. We are not confined like in state universities, which have [a] location-specific mandate." The scientists who focused on the state were all employed by a state agricultural university. The two scientists who responded that they work on specific conditions were both in Bihar. These responses are generally consistent with the mandates of the different research organizations.

When asked about how they accounted for climate and environment in their research, most scientists responded that they considered microclimatic factors in their research program, or that they hoped in the future to consider microclimate in the research. Scientists in northeastern India noted that there were several location-specific challenges there, including a lack of late-sown, short-duration varieties. Other challenges in the northeast included terminal heat stress, moisture stress, the need for more effective extension, and the need to recommend the right varieties to farmers and to take old varieties off the market.

Despite widespread acknowledgment of wide adaptation as a main goal of the wheat improvement program, many of the scientists expressed a desire to see more location- and environment-specific and breeding and research. Scientists reflected this in statements such as, "In future I'm hopeful that we are going to consider the microclimates at the district level," and "Because our zones are so large . . . if you propose a single variety for such a large area, it's not a good practice." Although varieties can be released at a subzonal level, the varietal testing system is set up to release varieties at the zonal level. Many scientists echoed calls to reevaluate the zonal orientation of the AICW&BIP.

The interview respondents made specific claims about why research for more specific climates could be beneficial. One research administrator stated that the "varietal development and testing program needs to be very careful today according to a smaller or different type of agroclimatic significant difference . . . maybe the number of varieties will be more . . . but the productivity, I believe personally, productivity may go up." Many scientists noted that only a few varieties tend to become popular. A scientist in northeast India said, "I would like to recommend the varieties suitable for specific conditions. Because the conditions differ very widely, if you go for fifteen kilometers, ten kilometers, there is [a] difference." A wheat quality scientist noted that "the zones that they have made are basically from a yield point of view or similarity of the agroclimate. But that doesn't suit for me as a quality person . . . we get location specificity in case of quality." Although some of these scientists were officially working on a zonal basis, they expressed hope that things would change in the future, leading to more research for specific conditions. Unfortunately, the structure of the AICW&BIP was not well aligned with this goal.

Scientists conveyed varying levels of dissatisfaction with the present system of breeding and testing, in which breeding is done under favorable conditions and then finished varieties are sent for multilocation testing. One researcher argued that the testing and release program should be opened to more actors than the public research system: "It will be better if we open our system of testing and releasing the varieties. . . . Our system of the coordinated trials [was developed] about twenty to thirty years before, when there [were] limited players in . . . seed production and the private players were not even present." But now that agribusiness has a significant footprint in India, the wheat system should "open to accommodate a large number of varieties" including submissions from private companies. On the topic of varietal adoption, they noted that the dominance of a single variety leaves wheat vulnerable to disease and abiotic stresses. "I presume it's suicidal to promote a single variety to be grown in a zone like the northwestern plain zone, where one variety at a particular point, say PBW343, CIMMYT's strain, was dominating over 9 to 10 million hectares, and that variety during its period of time was susceptible to yellow rust, but by chance it escaped during its period of dominance. . . . You are serving on [a] plate the condition for epidemic by growing a single variety." They noted further that scientists should even go "through other routes than the coordinated program, develop the strain and get it released to the farmers, so that diversity is created."

Based on my interviews, it appears that little research is conducted for specific conditions or locations (other than for favorable conditions). One scientist described the system of multilocation testing as "just looking into our selection of varieties and if something clicks, by chance, we send it." In other words, varieties are not bred for specific environmental concerns and are only assessed post hoc for suitability under rainfed or other conditions. Despite this, many scientists conveyed a desire to see more wheat-breeding and testing programs under diverse conditions. A scientist reflected, "To select for the area for which we are targeting, we should do our whole plant breeding in that area, meaning we should not [just] put our finished material in that area. . . . Then we will be able to get better yield. If we do the breeding for drought in irrigated areas . . . then we might get some of the genotypes but that might not be true for all." This perspective is all but prohibited from being put into action by the centralized nature of wheat breeding in India, which focuses on breeding for ideal conditions and does not involve farmers in the development or testing of varieties. For example, one scientist noted, "Whatever we breed at the research institution we give all required input, and then we say this is the yield potential, but the same may not be reflected in the farmers' field," particularly in Bihar, where irrigation is limited and lower socio-economic status farmers cannot afford much fertilizer. These concerns with the AICW&BIP echo the broader scientific discourse around target environments, and whether direct or indirect selection within a target environment is most efficient.[56]

A retired administrator summarized the historical focus on ideal environments justified by spillover effects: "With the introduction of the high-yielding varieties and the urgency to increase production, the focus was in the well-endowed, irrigated areas; very little attention to rainfed areas. Also, the general philosophy was that there are always spillover effects and therefore—and there are good examples where those spillover effects have affected—for example, some of the marginal varieties were bred for the irrigated areas, they found a place in the rainfed areas." Reflecting on the reason location-specific research is not pursued, the administrator stated:

> Addressing the specific needs of a region was never on the agenda. And this was true even for the agricultural universities. Because largely when we received international germplasm the flow was from the center to the ICAR institutes to the coordinated projects down the line. Now primarily a lot

of work they're doing under the local conditions, but . . . ultimately bigger
funding comes from the central sources. The specific focus on rainfed areas,
my own feeling is . . . limited. And at least definitely varieties are not bred
with those situations in mind. The kind of region specificity is one of the
major changes looking forward that should happen.

These statements, taken in whole, confirm that there are strong institutional structures and norms in Indian wheat research that favor wide
adaptation and centralized research over regional and decentralization.
Many higher-ranking scientists in the Indian wheat system promote
this status quo. But my interviews reveal that many scientists value context-specific research. They may feel a personal obligation to address the
problems of rainfed farming, or they may feel that the current system
has led to the release of inferior varieties. Despite this plurality of views,
however, these perspectives are marginalized within the centralized research system.

I also asked scientists whether they considered farm size or socioeconomic conditions in their research. Responses were mixed. Some
scientists took the view that seed technology is scale neutral and socioeconomics do not need to be considered. For example, "Since we are
targeting large areas, within that area farmers will be small, large . . .
we do not consider them individually." Some recognized that there are
socioeconomic issues related to technology adoption, but they left that
to agronomists, social scientists, agricultural extension, or the government to address. Scientists who responded that they did consider socioeconomics typically mentioned that some resource-poor farmers cannot
afford irrigation and fertilizer and machinery. One scientist responded
that small farmers have a greater preference for wheat quality and that in
contract farming, larger farmers derive more benefit. Altogether, the varied responses did not fall along disciplinary or regional lines but seemed
more related to the personal experience and philosophy of the scientist.

During the Green Revolution, plant breeding secured its role as the
most prestigious agricultural discipline, especially in India. Research
administrators tend to be breeders, and as can be expected, breeding
continues to be the focus of wheat research. These administrators have
benefited from the focus on wide adaptation because they gain prestige
by releasing a "blockbuster" variety—that is, a variety that has high yield
and wide adaptation. Thus, plant breeders tend to support the status
quo, focused on wide adaptation and research under ideal conditions.[57]
And because administrators decide on research goals, the system set up
during the Green Revolution continues today. Despite challenges to the
dogma of wide adaptation both within and outside of the research sys-

tem, the overall philosophy and structure of the Indian wheat research system is quite constant. I refer to wide adaptation as a dogma because in India, it is no longer a subject of study as it was in the post–Green Revolution period. Wide adaptation in India is a concept prescribed by elite researchers and one that scientists must accept in order to advance their careers. I call it a dogma, rather than a theory, because is remains a controversial topic in agricultural research. But to reject wide adaptation would be to reject the foundation of the current Indian wheat system that centralized around plant genetics. Importantly, I found that many Indian wheat scientists expressed dissatisfaction with the dogma of wide adaptation. Many (roughly half) of the scientists I interviewed expressed a desire to see more location- or condition-specific research. Despite the desires of Indian scientists to see more context-specific research, the Indian wheat research system steamrolls deviations from the norm, powered by the momentum of the Green Revolution.

BARRIERS TO CHANGE IN INDIAN WHEAT RESEARCH

So what keeps the prevailing research system in place, despite the diversity of views among India's wheat researchers? Institutional momentum—in other words, the tendency toward the status quo—plays a role. The status quo prevails for three reasons: (1) reliance on the linear model of innovation, (2) failures to create value from the public research sector, and (3) the "blame game." These are not problems that can fixed with more science and technology; they are norms and values. Each of these reasons is reinforced by the bureaucracy and centralization of Indian agricultural research, and supported by the actors who benefited from the consolidation of power during the Green Revolution.

THE LINEAR MODEL

Innovation is a buzzword in the development sector and it is often thought of as inherently positive.[58] We should, however, consider that innovations can have positive, negative, and ambiguous impacts on society.[59] Many scholars have pointed out that agricultural researchers typically adhere to the linear model of innovation.[60] This model is a common but misguided way to understand the process of innovation.[61] Under the linear model of innovation, scientists are not required to consider the social benefits or consequences of their research. Instead, they focus on producing basic science and assume that this will lead to socially desirable outcomes. This is like the transfer of technology and the "loading dock model" of research. In other words, "You take it out there, and you leave it on the loading dock and you say, there it is. And then you walk away and go back inside."[62] This model assumes a linear flow of knowledge and benefits from the scientist to the end user—in this case the farmer—with little or no contact or exchange of information. It also reinforces a sharp divide between scientists as producers, extension personnel as strictly educators, and farmers as passive consumers, thus precluding collaboration.[63]

The linear model is alive and well in Indian agricultural science and helps perpetuate the supremacy of plant breeding. New plant varieties are viewed as finished technologies that require little to no feedback from the farmer.[64] For a scientist, this model of technology development is easier and simpler than one that involves codevelopment with the user.

The centralized wheat research in northwest India also assumes that basic research from this area can be transferred equally well to other areas. Thus, there is no need for strong state and regional research if one believes in the linear model.[65] In the centralized, linear model many innovations remain "on the shelf" because scientists do not concern themselves with technology diffusion.[66] While basic research such as plant breeding is clearly important to agricultural development, what is the point if many new innovations are not reaching farmers? We should also not assume that innovations are a net positive to society, as many innovations favor early adopters and can cause inequality. Rather than following the linear model, in the next parts of this chapter I will use the term *innovation system* to describe the different actors, systems, and processes involved in the identification and promotion of new innovations.

PUBLIC VALUE FAILURES

A second barrier to a responsive agricultural innovation system in India is the public-sector orientation of wheat research, which is hampered by bureaucracy. Public sector research is often claimed to be a public good (as opposed to a private good). Public research systems are intended to focus on research that is less appealing to the private sector, such as low-value crops or innovations for smallholder farmers. In India, almost all wheat research is performed in the public sector—by the state agricultural universities, IARI, and the coordinated wheat program. Extension programs in India are also largely public sector, although many farmers get their information from the private sector, such as agrichemical dealers. But unlike the private sector, the Indian agricultural research system has few incentives to ensure that innovations are relevant and ultimately adopted.[67] This has led to a wheat research system that is highly bureaucratic, disciplinarily isolated, and unresponsive to farmer needs.

Since at least the 1970s, critics of India's agricultural research and policy have pointed out the large chasm between researchers and farmers.[68] In many developing countries, public agricultural research is the only source of formal innovations but "struggles to relate to its client base."[69] Kerala Agricultural University was once called "a burden on the State as it failed the farmers and the farming sector."[70] One challenge is that agricultural scientists tend to come from more elite backgrounds than farmers. There is not a culture of information exchange between scientists and farmers.[71] This is also due to the centralization of India's agricultural science, which enables little interactions between scientists and their stakeholders.[72]

There are few incentives for the system to meet either farmers' needs or work toward food and nutrition security. Just because a public sector intervenes to fill a market failure, this does not ensure a public good. We might call Indian wheat research as case in which "neither the market nor public sector provides goods and services required to achieve core public values," which is called a public value failure.[73] Public values would be reducing economic inequality and improving food security, rather than a goal such as raising agricultural GDP, which is a market value. Public value failures in science policy are difficult to correct because they often develop from systems with deeply entrenched interests, technocratic values, and bureaucracy. But the narrative that public research always creates public values is more appealing than a deep dive into the complexity of values, norms, and institutions of the system.

In the case of public value failures, public-private partnerships are often recommended to support inefficient public-sector activities. There is some evidence that privatized extension services are successful in India.[74] But the general institutional barrier facing the public research sector is the lack of end-user engagement. Despite a shift in global research toward engaging farmers, Indian agricultural research has strong social hierarchies that discourage consulting end users of technologies.[75] This ultimately results in technologies that are not adopted.

THE BLAME GAME

While India's wheat production has increased since the Green Revolution, many of its citizens are still far from food secure. Naturally, policymakers and researchers try to explain this disconnect between food production and food security. What results, however, is what Raina called the "blame game." Policymakers blame scientists, and scientists blame extension agents and farmers for not facilitating adoption of new technologies. Raina wrote: "When it comes to credit for food production, the agricultural technologies—the very varieties, irrigation, chemicals and pesticides, and the research organizations take it all. When it comes to the blame game—about persistence of rural poverty and hunger, child and adult malnutrition, environmental and social disruption, it is the other organizations and policies—the Food Corporation of India and the Public Distribution System, state and national level schemes for rural employment and poverty alleviation, input subsidies, rural credit, irrigation policy, international trade and the WTO [World Trade Organization], that are accused."[76] My interviews support this. Scientists often blamed agricultural extension for failing to transfer technologies to farmers. In a 1998 paper, agricultural researchers Mruthyunjaya and P. Ranjitha wrote, "Extension will necessarily have to adopt problem-solving approaches, and develop methods for transferring more site-specific information and technical knowledge."[77] While this is certainly true, it is easier to blame the extension system than to take more responsibility for ensuring technologies are well-fitted to local conditions.[78] S. Nagarajan, former director of IARI and the coordinated wheat program, wrote in 2005 that "national-level wheat production will change if only the states in NEPZ [Northeastern Plain Zone] and CZ [Central Zone] take their assigned responsibility with seriousness."[79] He blamed poor wheat production on the lack of "seriousness" in those areas and overlooked the systemic barriers to quality research and production there.

Scientists also blame farmers for not practicing "proper agronomy." Scientists classify wealthier farmers as "progressive" and "good" farmers,[80] blaming resource-constrained farmers for not adopting new technologies—especially recommended doses of fertilizer—and therefore having lower yields. My historical research confirms that "a chief complaint of all studies of the Green Revolution was that the Indian peasant did not apply anywhere close to the recommended dose of fertilizers on the dwarf plants."[81] I also found that scientists blamed intellectual property rights for creating a hostile system for scientists. For example, it was mentioned that unfinished varieties simply could not be given to farmers (a critical step in participatory breeding methods) due to intellectual property restrictions.

The blame game has two negative outcomes. First, of course, is that parties tend to absolve themselves of blame. Scientists can view their work as an isolated, value-neutral activity and rely on politicians and society to figure out how to best use innovations. This pattern is clear in the aftermath of the Green Revolution. Scientists and economists insisted that Green Revolution technologies were "scale neutral" but that inequality can be blamed on politics and society.[82] Second, all parties overwhelmingly fail to recognize that production is only one part of the food security equation. Scientists tend to focus on the "yield gap" between optimal and actual field conditions. I found Amartya Sen's work on the role of entitlements on food security seldom seriously discussed within agricultural research systems. Indian scientists and policymakers must critically reconsider the agricultural innovation system—including its goals, linkages, and outcomes—to improve food security in the country.

The three factors I describe here are not the only constraints to a more flexible and demand-driven agricultural innovation system in India but they are three that I find important to recognize and challenge. The next section examines some possible avenues for Indian agricultural science to move beyond maintaining the Green Revolution model and on to an innovation system that is both inclusive and useful to society.

PATHS TO CHANGING THE AGRICULTURAL RESEARCH PARADIGM IN INDIA

Addressing the barriers described above requires not just a paradigm shift within the scientific community but also what we might call "institutional innovations." Changing these systems is a slow process that can range from one to ten years for organizational changes to take hold, and 30 to 250 years for what Niels Röling calls "informal rules, customs, traditions, norms, beliefs" to change.[83] This endeavor would be magnitudes more difficult than the changes that occurred during the Green Revolution, because it runs against the momentum of the system. Some of these changes also challenge the values and beliefs held by scientists, administrators, and policymakers. While the Green Revolution centralized research and power, changing the Green Revolution paradigm requires decentralizing power, shifting from politics of control to politics of care, and systematically articulating and evaluating new science policy goals.

Innovation should not be viewed as a linear model that produces universal goods to society. We should instead view innovation as an iterative process that is driven by different stakeholder agendas and inputs. Farmers' adoption of agricultural innovations is not scale neutral but highly context dependent. Innovations are also not just technological; they are social and institutional as well. The Green Revolution occurred because of major institutional innovations around access to credit, inputs, and land reform. Even technological innovations should be viewed as just one part of the answer for smallholder farmers. Unfortunately, most agricultural research systems still emphasize technology, despite efforts to change this paradigm. Innovations, both technological and institutional, are critical to agricultural and food systems in India in the twenty-first century.[84] But in the face of climate change, poverty, and other challenges, innovation is not without risks. And innovation is not a politically neutral activity.

Innovation systems for smallholder agriculture and nutrition security must accommodate multiple "technological pathways" that can benefit different users in different environments.[85] This itself requires major institutional innovations within the Indian research system to adapt to a diversity of biophysical and socioeconomic systems rather than applying the one-size-fits-all approach. Marc Schut et al. described an example of carving "niches" for institutional innovations in agriculture.[86] If these niches are given proper resources and a champion, they can blossom without the overbearing influence of the current research system's norms, values, and incentives (which they call a "regime"). The authors described how "actors experiment with and organise around particular ideals, interests and, purposes" within these niches, which can lead to longer-term changes.[87] Along these lines, Rupert Sheldrake wrote, "Innovation will be most free when no particular orthodoxy achieves a monopoly of power, or of funding, and when scientific research in general, and agricultural research in particular, are carried out pragmatically, liberated from the dogmas of materialism, molecular triumphalism, and neoliberal capitalism, and placing a major emphasis on the building up of the quality of the soil and on sustainability."[88] An agricultural innovation system that allows more diversity can also lead to more diverse agroecosystems and diets.

The first step to a more inclusive agricultural innovation system is to decentralize the power of agricultural scientists. India's agricultural research system provides scientists with privilege and prestige that can make them resistant to reform. This power is ordained by the centralization of science, part of the movement known as "high modernity" that was prevalent during the 1950s and 1960s.[89] High modernity had a very straightforward application to agricultural science. Scientists, foundations, and governments believed that higher-yielding seeds would result in a more ordered peasant class. Scientists (especially plant breeders) had a powerful role that was solidified by the administrative changes in Green Revolution–era India.

During the modernization of agricultural science, scientists became even more removed from the messy world of farming and field research. With a change from local studies to a centralized approach, agricultural modernization codified scientific agriculture as the privileged form of knowledge (over indigenous knowledge, for example) and marginalized practitioners of all other methods.[90] The Green Revolution, however, failed to "impose standard top-down programmes and projects on diverse local realities where they do not fit or meet needs."[91] This is a classic outcome of high modernity—unruly subjects who refuse to fit the form prescribed to them by experts.[92]

Some researchers—especially social scientists—have attempted to address this failure through participatory agricultural research, engaging end users directly in the development or testing of innovations, thus giving them more power in the process. Participatory agricultural research became popular during the 1990s internationally but subsequently lost the interest and funding of international donors.[93] Some nongovernmental agricultural and international organizations in India use participatory methods, but it is not mainstream. The three barriers I have listed above help explain why participatory research never caught on in mainstream Indian agricultural research and why it waned in other contexts. Participation challenges the centralized system of agricultural research because it strengthens the role of the local and diverse while weakening control from the center. Participatory research lacks the uniformity of multilocation trials and scientists often have a difficult time changing their values around farmer participation. But participatory research can produce technologies that are more useful to farmers.

A more flexible, responsive agricultural innovation system also requires a shift from the politics of control to the politics of care.[94] In a strategy central to high modernity, agricultural scientists attempt to control the environment by minimizing the variation in fields and in yield responses. Where variation cannot be avoided, they employ broad categories such as rainfed versus irrigated lands that do not capture the nuance of location conditions.[95] The flaws of this approach are obvious, as agriculture is a dynamic biophysical and socioeconomic system that is complicated by the uncertain impacts of climate change. Care, on the other hand, acknowledges diversity and change as well as the need for specialized attention. The politics of care require time and effort to understand social and cultural realities and to develop innovations that are more sustainable.[96] Scientists and policymakers must recognize that technology-society interactions are highly context dependent. Top-down solutions have not improved livelihoods and food security because they ignore local complexity and the needs of farmers and consumers.

Now that we have identified some broad features of a more responsible agricultural innovation system, how do we implement them? If policymakers could convey the goals of agricultural innovation (beyond raising production), this would allow better targeting of programs and evaluating innovating proposals. Current innovation goals center on increasing production with some vague references to equity. Raina asks scientists themselves to "articulate the goals, demands, inadequacies, advances, and improvements needed in S&T to ensure delivery of development goods for the agricultural sector."[97] These are not just research goals, such as releasing a certain number of varieties with improved yield, but rather outcomes such as increased adoption of disease-resistant varieties. If these goals are articulated, scientists can assess their own success and better engage with policymakers.

A more fundamental institutional innovation, however, is to consider agriculture in a wider context. Agriculture is not an isolated system, and it has negative externalities such as pollution, groundwater depletion, and loss of biodiversity. I agree with Dominic Glover and Nigel Poole's call to assess "the quality of agri-food systems innovation, what pathways will be followed, and how the costs, risks and benefits will be distributed."[98] They propose an analytical framework to examine innovation proposals for food and nutrition security in India. Their framework reveals assumptions about seemingly simple "technological solutions" that may not be so simple.[99] Calling only for more agricultural production steamrolls other pathways to food and nutritional security and ignores tradeoffs with environmental sustainability.

ALTERNATIVE MODELS OF INNOVATION IN AGRICULTURE

Agricultural policy discussions in the public tend to coalesce around two paradigms: the "productivity" paradigm and the "agroecology" paradigm. The productivity paradigm focuses on increasing the yields of staple crops like cereals using a high-technology, high-input approach that is consistent with the Green Revolution approach. Agroecology advocates call for an approach to agricultural development that also considers social and ecological outcomes. While I criticize the productivity paradigm, and support aspects of the agroecology approach, the models I lay out in this section can be applied to both paradigms. These models allow agricultural research system to target specific groups of farmers, rather than develop one-size-fits-all solutions. The first model is called client-oriented research, and aims to improve the usefulness of agricultural innovations. The second is called the options by context model, and aims to provide farmers with a better range of choices to improve their productivity and/or livelihoods.

Client-oriented research considers farmers the client, or end user, of agricultural technologies. While client-oriented research may use participatory methods, it differs in that participatory research is a process and client-oriented research is an outcome.[100] Client-oriented research does not require technical or administrative complexity. Jonathan Harwood showed that Japan's agricultural development around 1880 to 1920 used this approach to develop high-yielding rice varieties.[101] Boru Douthwaite offers excellent examples of successful and failed client-oriented research in his 2002 book, *Enabling Innovation*. Using examples from different industries to show why top-down research often fails to have an impact, he argues that the most successful technologies engage their clients early and often to codevelop useful technologies.

Other agricultural scholars have expanded on Douthwaite's codevelopment model, drawing from market-based innovation models such as product development. James Sumberg and David Reece argue that "agricultural research might do well to learn the broader lessons of successful new product development."[102] Unlike Douthwaite, they find that not all types of technology development require user innovation, but all require an understanding of market forces and user needs even when users are not able to articulate those requirements clearly. They found that "some areas of innovation lend themselves more to participatory or co-development."[103] Take for example vaccines: vaccine development does not require much user engagement, but successful vaccine campaigns require understanding not only user interests but the infrastructure to deliver those vaccines. In agricultural contexts, Sumberg and Reece employ the new product development lens to assess users' "interest in obtaining benefits" beyond just their needs. The difference is that while users may have a particular requirement for a type of technology product, their overarching goals may be achieved by other, nontechnological pathways, such as greater access to nonfarm employment. Ultimately, agricultural research organizations will "require fundamental change in the culture and ethos" to adopt a new product development lens.[104] Laurens Klerkx and Cees Leeuwis found that "(historically derived) policy choices . . . often cause the norms and values of scientists to prevail over those of end-users."[105] Therefore, institutional changes are needed to promote client-oriented research.

The options by context model of research for development shares some features with client-oriented research, but it is more specifically focused on developing a better bundle of interventions that are appropriate to the contexts of smallholder farmers.[106] This fits the new product development lens in that "value of any particular technology or technical innovation will therefore depend on how well these benefits and resource requirements match the interests and available resources of potential users."[107] Katrien Descheemaeker et al. highlight how this model can be used to develop agricultural technologies.[108] They explain, "Climbing beans were initially identified by researchers as a 'best-bet' option for densely populated highland areas with small farm sizes, and technologies were further tailored to the local context during a co-design process. The process consisted of iterative cycles of design, implementation and evaluation with farmers."[109] This resulted in "about 20 promising, 'best-fit' options" that would allow farmers to choose the option that best fits their needs.[110] Fergus Sinclair and R. I. C. Coe lay out other research methods that can support the options by context model.[111] Again, this model would require significant institutional change in the agricultural research system.

CLOSING THE CHAPTER ON TRICKLE-DOWN AGRICULTURAL RESEARCH

The current system of agricultural research in India may be working for farmers who practice "proper agronomy," but that does not make it sustainable. Most farmers in India are smallholders operating under diverse conditions that may become even more variable as a consequence of climate change. The Green Revolution style of research glosses over this diversity and scientists use wide adaptation to justify catering to wealthy farmers rather than small farmers, rural people, or food security in general. We must dismantle the narrative of wide adaptation and the problematic assumptions that it encodes. To actually benefit smallholder farmers, innovation systems must be agile enough to target them directly. The trickle-down approach to research is not effective for smallholders and will inevitably benefit better-off farmers first. Pivoting to a more client-focused, decentralized research system requires a massive shift in the values and structure of Indian wheat research, and an overhaul in agricultural education, policies, and administration. Once accomplished, however, this will empower research focused on location-specific problems and agroclimatic variability that smallholder farmers face.

If India wants to support marginal and rainfed farmers, as it has claimed since independence, the current institutional arrangement of Indian wheat research requires serious modification. First, the government cannot continue to claim that agriculture is a "state subject" in India while centralizing research funds and activities. Removing all central support and leaving research to the states would clearly be disastrous and would lead to further bifurcation between wealthy and poor states.[112] But Indian administrators and policymakers must find a better means of allocating resources and power between regional and centralized agricultural research. Second, the current system of varietal development and testing should become more client and context oriented. Currently, varietal testing excludes farmer feedback and weeds out varieties suited to specific conditions. This system requires major changes to meet the needs of smallholder farmers.

More research is not always the answer to complex societal problems such as hunger. But if we do hold scientific progress as a public value, how do we ensure that science is both useful and just? Policymakers must decide what, exactly, the goals of continued investment in public agricultural research are. Is the goal simply increased cereal production? Improved nutrition? Higher value exports? Less imports? Interregional equity? Poverty reduction? Perhaps, as suggested by some researchers and development scholars, public policy should "support small family farms in either *moving up* to commercially oriented and profitable farming systems or *moving out* of agriculture to seek nonfarm employment opportunities."[113] This would mean focusing not on agriculture to reduce poverty but on increasing nonfarm employment opportunities. Once these goals are better articulated, the government should evaluate what parts of the current research system contribute to these goals and which require institutional change.

CHAPTER 5

CHALLENGES TO WIDE ADAPTATION IN INTERNATIONAL AGRICULTURAL RESEARCH

WHEAT IN NORTH AFRICA AND THE MIDDLE EAST AND MAIZE IN MEXICO

In this final chapter, we will step back into the late 1960s and early 1970s to look at three cases where wide adaptation was *not* an appropriate agricultural research strategy, despite its seeming success in Norman Borlaug's wheat program. These three cases are: a maize program in Mexico focused on smallholder farmers, a wheat research program in Turkey, and a new international center in the Middle East. All of these programs were connected to CIMMYT, the international successor of the Mexican Agricultural Program (MAP) that is headquartered in Mexico. These cases show how researchers in other political and agroclimatic contexts grappled with wide adaptation and critiques of the Green Revolution. They also show the limits of wide adaptation. Widely adapted wheat was most successful in areas with similar agroclimatic characteristics. No other crop has achieved the wide adaptation of spring wheat, yet it is still used as the model for international agricultural research.

By 1970 wide adaptation reigned supreme in international agricultural research. India had declared a wheat revolution and Borlaug was awarded the Nobel Peace Prize in 1970, propelling him to international fame. However, this coincided with criticisms of the Green Revolution, mostly from social scientists. Critics argued that Green Revolution technologies were inherently biased toward more commercial agriculture.

While theoretically any farmer could benefit from improved seeds, fertilizer, and irrigation, only farmers with more landholdings and capital could access credit and take the risk of adopting new technologies. Some agricultural commentators have called equity issues a "third generation" problem of the Green Revolution, which implies they could not have been foreseen.[1] However, Jonathan Harwood finds that "this argument does not stand up because the drawbacks of the technology did not become apparent only by the late 1960s. They were known and had been discussed in expert circles—in both global North and South—well beforehand."[2] The evidence I presented in the previous chapters supports this statement. Before jumping into the three cases, let us first examine how CIMMYT and other scientists responded to these critiques, as this provides important context for the three cases.

NEW DIRECTIONS FOR CIMMYT

In the late 1960s and early 1970s, CIMMYT and the Rockefeller Foundation expanded their research programs beyond Latin America and Asia and into the Middle East and North Africa. After the RF's Indian Agricultural Program ended, CIMMYT moved some of their and the RF's India staff to Turkey and to new international agricultural research centers. The Ford Foundation also moved key staff from India to North Africa and the Middle East. As these were mostly rainfed and dryland areas, the approach researchers had used in Mexico and India for wheat would not work here, and they needed a new one.

Agricultural administrators and researchers realized that the Green Revolution primarily benefited irrigated lands, despite Borlaug's claims that widely adapted varieties succeeded equally well in irrigated and rainfed areas. A CIMMYT report in 1971 noted, "Until the beginning of 1968, CIMMYT's wheat program concentrated almost exclusively on irrigated wheat . . . the varieties have been bred and produced primarily for the high producing, fertilizer-responsive irrigated conditions."[3] To reach new farmers in tropical and subtropical areas, CIMMYT realized that they needed to address the needs of rainfed agriculture. Surprisingly, Borlaug had earlier expressed the same thought even though it contradicted his idea of wide adaptation. He wrote to CIMMYT's director Edwin Wellhausen in late 1967, "We now talk about expanding the CIMMYT wheat work to the Anatolian Plateau, Iranian Plateau and the Mediterranean belt of North Africa and the Near East. To be successful there we must develop experts in wheat dry land farming techniques."[4] Borlaug now saw the utility in rainfed research, even though he and other RF staff told collaborators in the Middle East and India that breeding should focus only on irrigated conditions.

There was little disagreement that CIMMYT would need to shift directions in the 1970s. Part of the shift was pragmatic, due to a change in geographic focus. But CIMMYT was also responding to pressure from administrators and donors to address socioeconomic issues of equity. Administrators from the RF and the US Agency for International Development (USAID), among others, took notice of criticisms of the Green Revolution, notably criticisms of the uneven spread of technology and benefits. They started asking critical questions about whether research had improved the livelihoods of small and marginal farmers. CIMMYT added an economic studies unit in 1972, their first social science unit.[5] Over the next few decades CIMMYT shifted focus from targeted increases in food production, to agricultural development for economic growth, to improving food security and livelihoods.

Scientists, administrators, and donors all realized that the Green Revolution's benefits were limited to irrigated areas farmed by "larger, more commercially-minded, well-established farmers."[6] CIMMYT researchers recognized that despite large investments in research, "the problem of the small, non-commercial or semi-commercial farmer remains unchanged."[7] Even in Latin and South American countries where the RF had longstanding research networks, poorer farmers in highland areas were not benefiting from new technologies.[8] Donor organizations pushed agricultural researchers to focus on small and marginal farmers who were bypassed by the Green Revolution.

Broader trends in international development prompted changes in CIMMYT's focus. Robert McNamara became president of the World Bank in 1968 and led with a focus on global poverty reduction and equity.[9] McNamara "introduced into the Bank the 'people's basic needs' approach and shifted investments from a focus on the physical to the human part," as well as increasing overall investments from the Bank.[10] According to historian Michael Latham, McNamara had supported the modernization approach during his tenure as US secretary of defense, but he realized by the later 1960s that modernization had not led to more political stability, even in the United States.[11] McNamara believed that poverty reduction, more than infrastructure development and industrialization, would increase global stability. The World Bank, United Nations Development Program, and Food and Agriculture Organization of the United Nations (FAO) began focusing on poverty reduction and "a new ethos focusing on equity" as their primary goals.[12]

In 1969 McNamara proposed a new system of international agricultural research centers focused on achieving long-term outcomes. He felt that the global need for agricultural research and development had outstripped the capacity of US foundations.[13] This new system became the

Consultative Group on International Agricultural Research (now known as just CGIAR) in 1971. The CGIAR brought together CIMMYT and the International Rice Research Institute, as well as two other RF- and Ford Foundation–supported centers, the International Institute of Tropical Agriculture and the International Center for Tropical Agriculture. The CGIAR was financially supported primarily by the World Bank, the RF and the Ford Foundation, and member countries.[14]

The World Bank also started financially supporting CIMMYT directly in 1972.[15] New direction from the World Bank coincided with concerns from RF and Ford Foundation administrators about questions of equity and agricultural research for rainfed areas. The RF's director of agricultural sciences, Sterling Wortman, argued that "we must increasingly consider in our projections not only total national output, but as well, the number of people needing benefit," and former RF employee Ralph Cummings (then at USAID) wondered about "the problems of rainfed lands and small farmers," who were more numerous than irrigated farmers.[16] The challenge, however, was that there were few examples of technologies that exclusively benefited smallholder farmers. The CGIAR's 1978 Integrative Report brought up the challenge of reaching small farmers and pointed out that "even if such technology were developed," "large producers would very likely adopt whatever attributes of 'smallness' were needed to apply such technologies if it were to their financial advantage."[17] This challenge continues to plague agricultural development experts, who have largely ignored the examples of successful "peasant-friendly" technology development.[18]

By 1972, CIMMYT had broadened its main objective to include assisting the "development of food grain improvement programs . . . which will benefit the largest possible number of farmers, especially in developing countries."[19] While CIMMYT was still concerned with the "optimum production environment," a report stated that this was for research purposes only and "does not mean that CIMMYT is trying to benefit the irrigated, mechanized farmer, but only that CIMMYT provides a dependable first step for breeding and experiments."[20] CIMMYT would continue to focus on optimal production environments until the 1980s.[21]

CIMMYT's research philosophy has always been to produce widely adapted technologies. These technologies are then distributed to national research programs, which adapt the technologies to local conditions and coordinate with local extension to promote them. While CIMMYT provides some support to national systems, it is not involved in local breeding, testing, or extension. In another example of the "blame game," CIMMYT staff wrote in 1975 that "failure of national research programs to conduct trials on farmers' lands is one of the most common causes for error in research, and therefore, a major reason why farmers do not adopt the recommendations of the research service in some countries."[22] This is quite ironic given Borlaug's earlier messages and repeated pleas to the Indian wheat researchers to test wheat under high levels of fertilizer.

Wellhausen, freshly retired from CIMMYT, recognized the limits of CIMMYT's approach. He said that to spread technologies to more marginal areas, "we must remember that these delivery systems are going to be location specific and vary from region to region. What works in one place may not work in another."[23] This comment reflects both the diverse national systems and the actual limits of technologies due to socioeconomic and biophysical differences.

But CIMMYT researchers have been criticized for adopting the attitude that "center knows best" in agricultural research. This philosophy is also reflected in CIMMYT and the CGIAR's messaging around social inequity, which Edmund Oasa summarizes as, "Agricultural research cannot be fully accountable for the consequences of its work that are reserved for the wider framework of social forces."[24] These organizations converged around the message that agricultural technologies are "scale neutral" or "socially neutral" and that the outcomes are driven by social, economic, and political factors outside of the purview of research.[25]

Borlaug supported the scale-neutral narrative, stating in 1986, "Plant species are apolitical creatures. They cannot [be] coaxed to yield 10 times more on a small plot than they are capable of yielding on a larger tract of land, employing the same technology. The redress of social inequalities is a job that must be tackled largely by the politicians of the world, not the agricultural research community."[26] Green Revolution supporters claimed that there was no socioeconomic barrier to farmers adopting new seeds or other aspects of the technology package. In other words, that "the biological nature of the technology makes it scale-neutral."[27] Kapil Subramanian shows how Borlaug and the RF needed to emphasize the importance of seeds, rather than fertilizer or irrigation, to prove that the new technologies were scale neutral.[28] In this effort, "agricultural development was increasingly reduced to varietal development and adoption alone."[29]

Thus, the basic argument of wide adaptation across environments was expanded to include scalability across socioeconomic contexts. These assumptions underlie many current agricultural development efforts. I have discussed the flaws of the wide adaptation concept but will spend less time on scale neutrality. Scale neutrality was not uncritically adopted, but it did become part of the Green Revolution narrative. One of the India's Green Revolution architects, B. Sivaraman, gave a sharp-tongued speech in 1972 to the Indian Society of Agricultural Statistics, in which he commented that the promoters of scale neutrality felt "all that is required to remove the poverty of the small and marginal farmers is to provide them the credit and the inputs in time according to the routines prescribed by the scientists."[30] He then called this "the great illusion we appear to be unconsciously propagating."[31] Sivaraman detailed how "scientific agriculture" was too risky or unattainable for small and marginal farmers in India. Seeds are not scale neutral—inevitably, wealthier farmers benefit more from seed technologies because they can take on the risk of adopting new varieties. And in the case of high-yielding varieties from the Green Revolution, these seeds required fertilizers, irrigation, and precise agronomic management to reach their full yield potential.

CIMMYT and other organizations attempted to address issues of equity and small farmers in new research programs. Between 1968 and 1969 the RF, the Ford Foundation, and CIMMYT started three new programs: the Ford Foundation's Arid Lands Agricultural Development program, a CIMMYT/RF program in Turkey, and CIMMYT's Plan Puebla for maize. These three endeavors all showed that wide adaptation was not a tenable solution to raising yields in their target areas and crops (spring wheat, winter wheat, and maize, respectively). But while CIMMYT and similar research centers were locked into the dogma of wide adaptation, a new international center began under different circumstances. The International Center for Agricultural Research in the Dry Areas (ICARDA) would explicitly focus on drylands and use location-specific research methods. By briefly examining the Arid Lands Agricultural Development program, the CIMMYT/RF program in Turkey, and Plan Puebla, and the administrative discussions around them, we can better understand how wide adaptation came to be codified in certain research programs and not others. We can also see how agricultural scientists responded to direct critiques of their research and the legacy of these responses.

MAIZE IN MEXICO AND PLAN PUEBLA

We will go back to Mexico for the first case study of this chapter. The MAP wheat program was incredibly successful in irrigated central and northern Mexico. The maize program, however, was less successful. Farmers in central Mexico adopted new maize varieties at a much lower rate than wheat. By the mid-1960s, over 95 percent of Mexican farmers grew new wheat varieties, but only 13 percent planted new maize varieties.[32] The MAP's maize varieties, like wheat, were responsive to fertilizer and produced high yields under research conditions. So why were farmers not adopting new varieties of maize?

The MAP's initial work on maize focused on improving farming by testing imported varieties and adapting them to local conditions.[33] Scientists engaged in maize breeding because hybrid maize from the United States was not adapted to Mexican conditions. The MAP's early work on maize could be described as "peasant friendly," but the program goals shifted in the 1950s to concentrate on breeding maize for more commercial farmers.[34] For example, Wellhausen, one of the maize breeders, said in 1947 that "areas in which a surplus can be produced should be attacked first, the pure subsistence areas may be left to a long time in the future."[35] This was a precursor to the RF's strategy in wheat.

Despite the similar goals of the MAP's maize and wheat programs, the two crops faced vastly different agroclimatic and socioeconomic landscapes in Mexico. Most maize farmers were subsistence farmers on small tracts of land. They relied on rainfall for irrigation and were mostly located in the climatically variable central Mexico. The MAP's Delbert T. Myren estimated that there were "40 times more corn farmers than wheat farmers, and consequently 40 times as many decisionmakers to be reached with information about new production practices."[36] In 1962, twenty years after the MAP started, only 9.9 percent of Mexico's maize-growing area was irrigated.[37] Mexican wheat farmers, in contrast, were more compatible with the RF's wheat program. In the moderate desert climate of Sonora, wheat farmers held large tracts of land and had the socioeconomic ability to invest in new seeds, inputs, and production technologies. Eighty-nine percent of the wheat-growing areas in Mexico were irrigated in 1963 owing to government-sponsored irrigation projects, mostly in the northwest region.[38]

The MAP's research trajectory was also connected to the Mexican government's political agenda.[39] As Karin Matchett described, the idea of genetic improvement of maize varieties attracted interest from the leftist Mexican government under Lázaro Cárdenas, president from 1934 to 1940.[40] Cárdenas believed that scientific agriculture could improve the livelihoods of peasant farmers, and his government redistributed land according to the *ejido* system. Manuel Ávila Camacho, president from 1940 to 1946, shifted the discourse to overall production rather than peasant livelihood. Most government funding went to irrigated rather than rainfed crops under Camacho's successor, Miguel Alemán.

CIMMYT created Plan Puebla (sometimes called the Puebla Project) as the RF's first program to specifically target rainfed and small farmers.[41] Proposed by CIMMYT's director, Wellhausen, in 1966, Plan Puebla aimed to increase maize yields for farmers in the Puebla state, who did not have access to irrigation.[42] The project was meant not as a demonstration project but rather a learning process.[43] A conversation between Forrest F. Hill of the Ford Foundation and Borlaug in 1967 highlights the motivation for Plan Puebla. Hill wrote to Borlaug: "Sometime I want to talk to you about the problems of production on Ejidal [communal] lands. Del Myren recently sent me a paper for comments in which he talks about the need for incorporating Mexico's low-income farmers into her modern agricultural economy. If this cannot or is not being done in an environment as favorable as the Yaqui Valley," where CIMMYT was located, "it certainly is not going to be done under less favorable circumstances."[44] Hill meant that if CIMMYT could not reach small farmers in this area with adequate rainfall, then they were less likely to succeed in other areas. Puebla faced reliable rainfall and was considered relatively favorable for maize production.

CIMMYT researchers worked with the National College of Agriculture of Chapingo and smallholder maize farmers in the state of Puebla to enact the plan in 1967.[45] Plan Puebla aimed to dramatically increase maize yields "by adapting existing technologies to ecologically specific growing conditions" of the Puebla state.[46] Scientists would first determine which maize varieties were a good fit for the region, and then breed varieties for those environments. Researchers promoted the package of Green Revolution technology, including fertilizers, improved seed, and agronomic management.

But after one year, researchers realized that their "improved" varieties had no yield advantage over the local varieties.[47] Scientists then focused on developing the best agronomic practices for the region, primarily higher fertilizer rates and planting density. Ultimately, Plan Puebla did substantially increase the yields of farmers in the Puebla region in a short period of time. Researchers noticed two interesting results: farmers very often did not adopt the "complete package" of recommended practices, and farmers found that local seed was more responsive to fertilizer than CIMMYT varieties were.[48] CIMMYT's economic analysis of Plan Puebla also found that farmers could improve their net income by improving agronomic management (without adopting new varieties), especially farmers who had access to credit.[49]

But by 1973 CIMMYT discontinued the Plan Puebla, to the disappointment of some RF advisors.[50] A 1972 report by Myren, then at USAID, outlined some of Plan Puebla's constraints. One of the main findings of Plan Puebla was that CIMMYT's maize varieties were not competitive with local varieties, which were selected over time by Mexican maize farmers for their adaptability to stress conditions. Myren wrote that "in contrast to the experience with the 'green revolution' in wheat," for maize, "identifying improved germ plasm for rainfed production is much more complex than it is for irrigated conditions. It also appears that up to now most breeding programs have focused on selecting material for optimum moisture conditions."[51] He indicated that the high yield potential of CIMMYT maize varieties was the result of a "good fit between the characteristics of that variety and the particular environment" rather than an inherent adaptability.[52]

Myren concluded that contrary to the of concept breeding for wide adaptation, "in the Puebla area the variability in soil and climate made it impossible to select one central spot that would give results applicable to the whole area."[53] This unfortunately led CIMMYT to decide "that this project is too far afield from its main interest in plant breeding," and the program was ended.[54] Plan Puebla was subsequently adopted by the Mexican government as a program for rainfed agriculture.[55] Plan Puebla led political scientist Kenneth Dahlberg to wonder "what the shape of agricultural research in those developing countries influenced primarily by the Rockefeller approach would be today if Project Puebla had been started in 1941 instead of the original project for increasing production on irrigated soils through specialized seed technologies."[56]

A 1969 CIMMYT report described Plan Puebla as "CIMMYT's answer to the question: how can the large traditional-agriculture sector be transformed into modern farming?" and claimed that if the plan succeeded, scientists could "bring the green revolution to thousands who have heard of it or seen its benefits—in the fields of the large farmer and on state-owned lands."[57] Thus, Plan Puebla was a study of whether the benefits of the Green Revolution could be extended to marginal lands. This is obvious in the follow-up reports titled "Plan Puebla: Transferable and Generalizable?" and "The Lessons of Puebla and the Potential of the Puebla Approach."[58] But instead, Plan Puebla demonstrated that plant breeding focused on ideal conditions did not necessarily spill over to benefit small, rainfed farms. Plan Puebla showed that at least for maize in Mexico, scientists needed to focus on more specific agroclimatic and socioeconomic contexts.

THE ROCKEFELLER FOUNDATION AND CIMMYT'S WHEAT PROGRAM IN TURKEY

In the immediate aftermath of India's wheat revolution, the Rockefeller Foundation and CIMMYT started a wheat program in Turkey. Wheat is an important crop in Turkey, but yields were persistently low in Turkey's large rainfed areas. Turkish wheat production had increased from 1958 to 1967 as a result of improved agronomic practices, but only using local varieties. Turkish wheat production faced problems of overall low yields, low moisture, and plant diseases such as rusts and septoria (a fungal infection). In 1966, a group of "progressive farmers" purchased some Sonora 64 seeds from Mexico.[59] The Mexican varieties quickly spread in the coastal areas of Turkey, which were well irrigated, but not to the interior part of the country. CIMMYT hoped to extend the Green Revolution to Turkey's wheat-growing interior, known as the Anatolian Plateau.

In 1969 the government of Turkey asked the RF for assistance with their national wheat improvement program, which started in 1967.[60] The RF recruited wheat scientists from CIMMYT and Oregon State University (OSU) to move to Turkey, and USAID provided additional financial support.[61] The bilateral agricultural program between the RF and Turkey aimed to train Turkish agricultural scientists and increase Turkish wheat production. CIMMYT's board of directors also had broader hopes, stating that "Turkey itself represents an important wheat production area of the world" and "the program could serve as a CIMMYT 'outpost' for . . . a larger area."[62] As CIMMYT had just formed in 1966, it was figuring out its position in international agricultural research.

Turkey is somewhat unusual in that it grows spring, winter, and facultative (having qualities of both winter and spring) bread wheats thanks to its diverse agroclimate. This is important because the wheat varieties released by CIMMYT thus far were only spring wheats. Winter wheats differ from spring wheats in that they need a cold period, while spring wheats are damaged by cold temperatures. The coastal wheat region of Turkey has a moderate climate, is mostly irrigated, and grows spring wheat. These conditions differ greatly from the Anatolian Plateau, which contains a greater share of Turkey's wheat production by area but has lower yields. The Anatolian Plateau is mostly rainfed and, because of its climate, can grow only facultative and winter wheats. Because "most of the winter wheats introduced into Turkey are not well adapted," CIMMYT initiated an international winter wheat screening nursery in Turkey in 1972.[63] Even winter wheat varieties from the US Pacific Northwest, which had similar agroclimatic conditions to the Anatolian Plateau, were not a good fit for this region. There were also no commercially grown winter wheats with resistance to stripe rust.[64] By engaging in Turkish wheat production, CIMMYT pushed beyond the agroclimatic constraints of its program on irrigated spring wheat.

The RF chose Bill C. Wright as the project leader for Turkey, transferring him from India in 1970. Wright codirected the Turkey program with Ahmet Demirlicakmak, a prominent Turkish scientist. CIMMYT then brought on Arthur Klatt in 1971 as a wheat breeder, and J. M. Prescott as a pathologist. Prescott had been stationed in India before coming to Turkey, and Klatt with CIMMYT in Mexico.[65] CIMMYT's Turkish wheat program shared similarities with the Indian wheat program beyond its staff. The initial philosophy and methods of the program were based on coordinating research by agronomic zones.[66] The program also aimed, according to Klatt, to "test our materials under varying environmental conditions . . . and also identify lines with broad adaptation."[67] OSU scientists Warren Kronstad and Tom Jackson consulted and helped bring farmers and extension agents from Oregon to Turkey to exchange ideas and practices.[68] By 1974 four additional international staff joined the RF/CIMMYT program in Turkey: Floyd Bolton, an agronomist appointed by USAID; Michael Lindstrom, a soil scientist; Charles K. Mann, an economist; and Dwight Finfrock, who was appointed to help develop the extension stations.

The RF/CIMMYT program in Turkey coincided with criticisms of the Green Revolution. One of these criticisms was that Green Revolution wheat varieties had less-stable yields over time than local varieties.[69] In other words, critics claimed that modern varieties could have good yields under good years but had greater yield losses than local varieties in bad years. CIMMYT countered this by making "yield stability" one of their core goals in the 1970s. Yield stability meant stability over time, not just space, and scientists promoted it as complementary to wide adaptation. CIMMYT's economist in Turkey, Mann, was somewhat skeptical of claims around yield stability. He wrote to a colleague, "Since for a long time people have talked of greater yield stability with improved technology, it seemed important to distinguish between higher average yields and lower year-to-year variation."[70] For example, results of the international wheat nurseries were typically presented as the average yields of each variety, rather than the yield variability over time. Mann pondered whether "because of the interactions between the package and moisture, annual variation may actually become greater rather than smaller" in CIMMYT's varieties, an argument other critics have made through empirical data.[71] Many scholars have pointed out that a farmer would not care about high average yield across locations, since they are only farming in one place. But they care about stability of yield over time, because this lowers their risks during bad weather years.[72] The question of yield stability remained a controversial topic at CIMMYT throughout the 1970s and 1980s.

Klatt, the CIMMYT wheat breeder in Turkey, spoke at a 1973 CIMMYT wheat symposium about yield stability. He opened by saying, "Since the science of varietal improvement was initiated, we have worked to increase maximum yield potential. Today, I would like to discuss this as well as a new topic, stabilizing minimum yield levels."[73] Klatt noted that it was more difficult to obtain a stable, minimum yield in winter

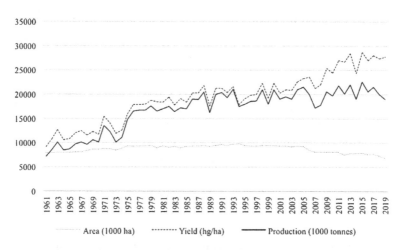

FIGURE 5.1. Turkey's wheat production in 1000 tonnes, wheat area harvested in 1000 hectares (ha), and wheat yield in hectograms per hectare (hg/ha) from 1961 to 2019. Data obtained from Food and Agriculture Organization of the United Nations, *FAOSTAT.*

wheat varieties in Turkey. He explained, "the main reason is the climatic instability of the winter wheat growing regions. Most winter wheat areas are characterized by large annual fluctuations in temperatures and precipitation and also large monthly fluctuations," and that scientists should breed winter wheats for yield stability.[74] CIMMYT scientists hoped that improved varieties would increase yield stability in Turkey.

In Turkey, CIMMYT started using what is called "conservation agriculture," a method to reduce soil erosion and conserve soil moisture by minimizing soil disturbance. This practice is currently widely promoted in agricultural development. RF, CIMMYT, OSU, and Turkish researchers experimented with different agronomic techniques that would help wheat farmers retain soil moisture, prevent soil erosion, and control weeds. Some of the conservation agriculture techniques were imported from OSU and modified. The US Pacific Northwest climate is similar to the Anatolian Plateau in terms of rainfall, but differs in temperature and soil type.[75] The scientists first experimented with variables like tillage, planting dates, and chemical inputs on research plots. Then they tested the best approaches in farmers' fields, which they called "adaptive research."[76] Agricultural extension then demonstrated the recommended techniques, and Mann calculated the improved profit from the agronomic techniques. CIMMYT researchers found that, similar to what had happened with Plan Puebla, farmers tended to adopt only parts of the "package," and Mann wrote that "there is also value in formulating a model to explain what farmers are actually doing as opposed to what it is recommended they do."[77] He emphasized that the success of the Turkey program came from the strong interdisciplinary team consisting of both agronomists and economists.

Ultimately, Turkish wheat production has greatly increased since the RF, OSU, and CIMMYT's involvement (fig. 5.1). Wheat production almost doubled from 1970 to 1980.[78] But except for the coastal areas, where semidwarf varieties were adopted, most of the wheat yield improvement came from agronomic improvements, not new wheat varieties.[79] The director of CIMMYT's wheat program at the time, Byrd C. Curtis, stated in 1980 that "this impressive increase resulted primarily from the application of water-conserving cultural practices on the Anatolian Plateau. Varietal changes contributed little to this gain."[80] In this respect, the Turkish wheat differed greatly from the RF's program in India. It also disrupted the notion that farmers needed improved varieties and genetic breakthroughs to improve wheat production.

In 1976 the RF/CIMMYT Turkey program started reducing staff because Turkish scientists now had the necessary experience to take the lead. By 1976 CIMMYT also decided to phase out countrywide programs and focus on providing regional assistance, working with the newly formed ICARDA. The CIMMYT program in Turkey officially ended in 1982. In 1986 CIMMYT and the government of Turkey started the International Winter Wheat Improvement Program, which was centered in Turkey, and aimed at developing facultative and winter wheat varieties for the region.[81]

The RF/CIMMYT experience in Turkey was very different from Mexico and India. Scientists realized early in the program that there were no improved varieties for the rainfed Anatolian Plateau, and even if there were, they likely would not perform substantially better than local varieties. Working as a small, interdisciplinary team and engaging local scientists and farmers, the foreign scientists focused on cost-effective agronomic techniques to help raise wheat yields. Even these techniques could not be directly imported from a similar agroclimatic region, but required adaptive research to fine-tune them to local conditions. In the passing years, Turkish and international scientists have developed improved varieties for interior Turkey that have higher yields and yield stability. But *semidwarf* winter wheat is not grown in Turkey's rainfed areas because it is not suitable to the dry, resource-limited environment there.[82] Improved varieties are also not necessarily reaching farmers: 49 percent of Turkey's wheat area is planted under varieties that were released before 1995.[83] Turkey is an example of how researchers specifically targeted socioeconomic and agroclimatic populations. In this case, researchers deviated from the expected Green Revolution package and were still successful in improving wheat production.

FOUNDATIONS IN THE MIDDLE EAST

The RF and CIMMYT planned to intensify their research in North Africa and the Middle East (sometimes referred to as the Near and Middle East) in the late 1960s. While the RF was involved in this region from 1960 onward and the FAO had worked there even earlier, there were no permanent international agricultural research networks or hubs in this region. Efforts to establish an international research center in the Middle East in the 1960s contended with criticisms of the Green Revolution approach, and also challenged CIMMYT's focus on varietal improvement and wide adaption. The newly formed ICARDA chose a very different approach from CIMMYT. This created anxiety and resistance among some CIMMYT scientists, who felt that ICARDA was a competitor rather than a potential collaborator. If wheat was so widely adapted, then why did the Middle East need its own research center to work on the same crop?

Starting in 1960, the RF had brought wheat scientists from North Africa and the Middle East to Mexico for training in partnership with the FAO. The RF also had established several wheat nurseries around the region. Rockefeller Foundation officials had hoped back in 1963 that "the wheat research work in India might serve as the eastern anchor for cooperative activities through the Near East countries."[84] After their experience with wheat in India, CIMMYT felt "there is much interest in revolutionizing wheat production in Africa and the Near and Middle East. . . . The knowledge, materials and trainees developed by the Mexican-CIMMYT wheat program have been decisive catalysts in provoking wheat production revolutions in several countries."[85] Despite wanting to extend the Green Revolution into the Middle East, CIMMYT and the Ford and Rockefeller Foundations came upon new challenges with dryland wheat. This led to research shifts and to new institutional arrangements. North Africa and the Middle East grew primarily rainfed spring wheat as well as durum wheat. Although Borlaug had previously claimed that his widely adapted varieties were superior to local varieties, it became clear that this technology could not simply be transposed into rainfed and dryland areas in the Middle East.

The RF sent scientists Elvin Stakman and John Gibler to the Middle East in the fall of 1966 to prepare a proposal on a regional wheat program there.[86] The RF was "poised to initiate a wheat research center, involving a 10-year commitment of around $4 million" in Lebanon with wheat breeder Gibler as its director.[87] Then war broke out in Lebanon in June 1967 and the RF withdrew completely. But the RF and CIMMYT continued to expand in the Middle East and North Africa region in the late 1960s. In addition to the Turkey program, CIMMYT partnered with US-AID and the Ford Foundation to start a regional wheat production campaign in Tunisia and Morocco in 1968, stationing five scientists there by 1969.[88] CIMMYT also worked to improve wheat production in Egypt. New financial support from USAID in 1969 facilitated this expansion.[89]

The Ford Foundation also had plans to work on agriculture in the Middle East around this time. Hill, who was then the Ford Foundation's vice-president for overseas development, had suggested a regional agricultural center in the Middle East in the early 1960s, and his successor, David E. Bell, continued that path. The foundation already had an office in Beirut, and despite the 1967 civil war, it established a research center called the Arid Lands Agricultural Development (ALAD) Program in Lebanon. ALAD was a short-lived research program specifically focused on North Africa and the Middle East. It was not the ideal time to start a research program in the Middle East, but Hugh Walker, a Ford Foundation representative in Beirut, lobbied for the creation of ALAD. It actually started before it was approved in the foundation's main office in New York, but officially, the Ford Foundation initiated ALAD in February 1968. Walker was the program's first director, and then Robert Havener took over in 1971. ALAD continued until December 1976.

The Ford Foundation did not want to create another CIMMYT or International Rice Research Institute (IRRI). Instead, the foundation followed its own model of outreach, extension, and community development for ALAD.[90] ALAD was not a centralized research center like CIMMYT but rather a decentralized network for testing varieties using adaptive research.[91] ALAD focused on agriculture issues of North Africa and the Middle East, from "Afghanistan to Morocco and Turkey to Sudan and Ethiopia."[92] The Ford Foundation noted that "the national agricultural institutions in the region lacked the capability or interest in testing, evaluating and modifying them [CIMMYT varieties] to fit local conditions."[93] Therefore, ALAD at first focused on extending "the benefits of the green revolution to the Middle East through adaptive research."[94] Initially, "ALAD functioned essentially as a relay station for CIMMYT and other 'green-revolution' technology and training centers."[95] Because of the success of Borlaug's wheat in other regions, administrators hoped that ALAD would "lead with wheat" and then move to other crops in the region.[96] ALAD also focused on pulses, forages, and sheep production.

In an ironic choice for a program specifically focused on arid lands, ALAD first focused on irrigated crops, especially irrigated wheat, because training and materials were readily available. But a review of the program in 1971 brought the "realization that the region demanded location specific research" for disease, rainfed agriculture, water management, and so on.[97] ALAD later expanded to include coarse, rainfed grains such as sorghum and millet. The demand for location-specific, and not just adaptive, research would also carry on to ALAD's successor, ICARDA. ALAD also took a more collaborative approach to plant breeding with farmers in comparison with CIMMYT. Joseph Remenyi, an economist who conducted an evaluation of ALAD, wrote that it was innovative in its nursery programs. He wrote, "Responsibility for selection of well adapted high yielding varieties was not kept at the ARI [Agricultural Research Institute in Lebanon] for ALAD staff. They sent their cooperators seeds of early generation materials. . . . The cooperators would then need to select out the best varieties, making them real partners in the regional breeding and improvement program."[98] This was quite a different program from traditional plant breeding, in which varieties are developed only under controlled conditions, and farmers and cooperators are not involved until they receive "finished" varieties. Researchers now call this approach participatory varietal selection, and ALAD was one of the first international research programs to use it.

The RF and CIMMYT became more directly involved in ALAD a few years after it was established. Borlaug was appointed as an advisor to ALAD. In 1971 the RF provided Leland R. House, who had worked with the coordinated maize program in India, as staff for ALAD. Eugene Saari, who worked with the Ford Foundation and CIMMYT in India, joined ALAD from 1973 to 1975 and then remained affiliated with CIMMYT until his retirement.[99] Saari had helped establish a regional disease screening nursery in India, which was extended to the Middle East and North Africa while he worked at ALAD.[100] Consultants from the Ford Foundation, the RF, the FAO, and CIMMYT established a set of unified wheat nurseries for the region in 1971, including the Regional Disease and Insect Screening Nursery.[101] Collaboration between the Ford Foundation, RF, and CIMMYT continued throughout the 1970s in North Africa and the Middle East when ICARDA was formed. ALAD also had an influence on CIMMYT: ALAD's director, Havener, become the third director general of CIMMYT from 1978 to 1985.

ALAD set the stage for another international program that still exists today. The work of the RF, Ford Foundation, and CIMMYT in North Africa and the Middle East eventually formed ICARDA, an international research center focused on dryland agriculture in the region. ICARDA was formed in 1977 and built on ALAD's infrastructure and staff after ALAD ended in 1976. The lessons learned from ALAD also shaped early priorities of ICARDA, although there were differences between the two organizations. Remenyi's review of ALAD found that "it was inevitable that the majority of its resources were optimal in environments that could most benefit from these [Green Revolution] technologies, typically the irrigated wheat areas. Little in the way of rainfed wheats was to be found."[102] But ICARDA was determined to be "a dryland agriculture successor to ALAD."[103]

While ALAD was still active in late 1971, a "proposal for regional coordination of the Mediterranean and near east cereal programs" was presented to the CIMMYT board of directors' executive committee.[104] The 1971 proposal noted existing collaboration between the RF, Ford Foundation, and USAID in the Middle East region. The CIMMYT board of directors was reportedly "enthusiastic" to move forward.[105] The next important step was to get support from the newly formed CGIAR for a dryland research center in the Middle East. The CGIAR was moving away from foundation-funded research centers like CIMMYT and IRRI and toward a more coordinated approach. But in 1972 the CGIAR's technical advisory committee noted that North Africa and the Middle East "were receiving insufficient international support in agricultural research by comparison with other developing regions."[106] While the FAO, CIMMYT, RF, and Ford Foundation were active in this region, various actors wanted an international research center there analogous to CIMMYT.

In 1973 the CGIAR's technical advisory committee commissioned a team led by Dunstan Skilbeck, of London University, to investigate the region and recommend a solution.[107] The team produced a report, *Research Review Mission to the Near East and North Africa*, which proposed an international research center there.[108] This report recommended that a new center focus on "research on barley and lentils; adaptive work on wheat and some other cereals (such as sorghum, pulses, and rice) on the basis of materials supplied by CIMMYT, ICRISAT [International Crops Research Institute for the Semi-Arid Tropics], and IRRI; and research on farming systems of dry areas, including both crops and livestock."[109] It noted that rainfed bread wheat, winter wheat, barley, and "the durum wheats, for which the Region is particularly noted, have until recently been neglected, and require much more attention."[110] Durum wheats, which are used for pastas, couscous, and flatbreads, are grown under mostly rainfed conditions and are native to the Middle East.

Although CIMMYT's board of directors was initially "enthusiastic" about a research center in the Middle East, CIMMYT's wheat scientists, including Borlaug, R. Glenn Anderson, and Keith Finlay, strongly resisted the idea of another center working on bread wheat, durum wheat, or barley. CIMMYT had started working on durum wheat in 1968 when it expanded into North Africa and the Middle East, and started working on barley in 1972.[111] Further, by 1973 CIMMYT already had staff working on wheat in Turkey, Lebanon, Tunisia, Algeria, and Morocco; another wheat center seemed redundant.[112]

Borlaug, Anderson, Finlay, and CIMMYT's director, Haldore Hanson, wrote a strongly worded report to rebut the CGIAR report. Because CIMMYT's wheats, and presumably barley, were widely adapted, there was no need for a regional center, they argued. They wrote that CIMMYT's wheat varieties were not specific to Mexico; they "might more correctly be called 'World Network Wheats.'"[113] Borlaug and his coauthors also resisted the idea of a region-specific research center, writing that "*international institutes can best be managed on a global basis, not a regional basis.*"[114] They also wrote that "wide adaptability cannot be achieved if progeny are selected and tested within one climatic zone," meaning that the varieties developed in the Middle East would be restricted to that region.[115] They believed that CIMMYT should keep its global mandate for wheat, durum, and barley, and that "no second CIMMYT needs to be created."[116]

Despite CIMMYT's objections, the CGIAR went ahead with plans for ICARDA. David Hopper, of the International Development Research Centre of Canada (and formerly of the Ford Foundation in India), led a subcommittee in 1975 to establish the center.[117] The original plan was to have three stations in the region under different agroecological conditions. The CGIAR was working on the legal proceedings to start cooperative centers in Lebanon, Syria, and Iran, but Lebanon experienced a civil war in 1975, and in 1978 riots broke out in Iran.[118] So ICARDA was headquartered in Aleppo, Syria. ICARDA was officially formed in 1976, with funds from Canada and sixteen total donor countries and institutions.[119] Most of ALAD's staff was transferred to ICARDA when it began operating on January 1, 1977.[120]

In line with the concerns about equity at the time, ICARDA had a broad goal of improving "the social and economic well-being of rural people in the region it serves."[121] The CGIAR's technical advisory committee decided to give ICARDA a research mandate for bread wheat, barley, and legumes, although the Ford Foundation disagreed with giving ICARDA an international mandate for any specific crop. CIMMYT would continue its durum program, to be discussed with ICARDA later. ICARDA's initial research objectives were "to improve the level and stability of production of the staple food crops of the region," especially for rainfed conditions.[122]

ICARDA was in many ways an extension of CIMMYT model, but its research program explicitly included rainfed areas and considered farmers' socioeconomic factors. Despite its similarities to CIMMYT, IRRI, and other international research centers, ICARDA was a deliberate shift away from the Green Revolution focus on only irrigated regions. The center promoted agronomy and socioeconomic analysis over strict germplasm improvement, although they started plant breeding in 1981.[123] ICARDA grappled with this new direction, as M. B. Russell, a Ford Foundation consultant, asked, "Should ICARDA devote a major part of its effort and resources to programs which will not yield an identifiable production such as an IR-8"—one of IRRI's first released rice varieties—"or dwarf wheat?"[124] He continued, "If so[,] what criteria should be used to measure Program effectiveness? Will such criteria be accepted as valid by donor agencies and the general public?"[125] These questions have perpetually plagued the CGIAR research centers. Researchers regularly use farmer adoption of varieties as a yardstick to measure impact, rather than examining socioeconomic impacts such as economic returns or livelihood improvements.

Over the next decade, CIMMYT and ICARDA wrestled with the mandates over durum and barley. The CGIAR asked their technical advisory committee to investigate the division of durum and barley between them after a routine review of the centers in 1983. With some mediation by the technical advisory committee, the centers decided that CIMMYT would have global mandate for durum breeding, but ICARDA would work on regional breeding for durum. ICARDA would also expand their barley breeding. The technical advisory committee found that CIMMYT and ICARDA's wheat strategies were complementary rather than competitive. On ICARDA, they wrote:

> ICARDA's targets in cereal improvement are the low to medium rainfall areas of the region, where the bulk of the durum wheat and barley is grown. Its breeding strategy, essentially, involves screening of the parental materials for tolerance to specific stresses prevailing in the region, crossing the superior lines thus identified, and then exposing early generation populations to these stress environments to allow identification and selection of superior gene combinations under these conditions. In later generations the material is evaluated at carefully selected multilocation testing sites to finally select entries for the International Testing Nurseries to be supplied to the national programs.[126]

This was quite a different strategy from CIMMYT's, although CIMMYT was becoming more engaged in research for water- and heat-stressed environments. On CIMMYT, the technical advisory committee wrote:

> CIMMYT's breeding philosophy is targeted towards a broader front and essentially involves initial selection under higher potential yield conditions coupled with shuttle breeding, permitting two breeding and selection cycles

per year, using sites differing widely in environmental conditions, disease spectrum and photo-thermal conditions. This procedure, besides being faster, allows CIMMYT to incorporate wide adaptation into the germplasm. The germplasm so developed is evaluated in all the major small grain growing areas across a wide range of climate zones and diverse soil conditions. Varieties from this program have performed well over a wide range of agronomic and climatic conditions.[127]

These two approaches exemplify the ideological divide between CIMMYT and ICARDA, as well as a broader divide in the plant-breeding community over breeding and testing under stressed versus ideal conditions.

Ultimately, both CIMMYT and ICARDA wheat varieties have found a place in North Africa and the Middle East. In a CIMMYT report, researchers estimated that over 80 percent of spring bread and durum wheat in this region was derived from CIMMYT and ICARDA lines, although winter and facultative wheat derived from the international centers had a lower adoption rate, 70 percent.[128] ICARDA's durum varieties Waha and Om Rabi 1 became popular in Algeria and Morocco, respectively.[129] Thirty percent of winter or facultative types of wheat are still landraces, though, which means that modern varieties are not fully adapted to these conditions, and/or modern varieties are not reaching some farmers. ICARDA, like ALAD, has engaged in participatory plant breeding with farmers, which has been efficient at selecting varieties that farmers prefer.[130]

The experiences of ALAD and ICARDA show the evolution of perspectives on wheat research in North Africa and the Middle East. ALAD was envisioned as an adaptive research center for irrigated wheat, while its successor, ICARDA, would focus on agricultural production for the specific conditions of the region. The different reactions of CIMMYT to ALAD versus ICARDA shows how wide adaptation was ideologically ingrained in the mission of CIMMYT. ALAD was not a threat to wide adaptation, while ICARDA challenged it.

The three cases in this chapter show how international wheat researchers responded to critiques of the Green Revolution starting in the late 1960s. In Plan Puebla, CIMMYT wanted to show that rainfed agriculture also benefited from Green Revolution innovations. The Rockefeller and Ford Foundations and CIMMYT worked in Turkey and created ALAD initially to transfer Green Revolution lessons to other geographies. In each case, however, the intended approach of technology transfer was unsuccessful. Instead, researchers tested new approaches to agricultural development based on agronomic improvements, location-specific research, and farmer participation. Unfortunately, the CGIAR has not developed the institutional-learning capacity to learn from these cases.[131] Rather than learning from the early lessons of Plan Puebla and Turkey, CIMMYT continued a "narrow focus on seeds—and seeds which required either irrigation or abundant rainfall, plus expensive inputs—rather than on how to increase the welfare of rural dryland peasants."[132] Agronomy has always been second to plant breeding at CIMMYT and many of the international research centers.

LESSONS FROM INTERNATIONAL RESEARCH AFTER THE GREEN REVOLUTION

There are some lessons we can glean from these cases and apply to current agricultural research. One lesson is that agricultural development cannot be reduced to technology development. While some might imagine that farmers can "plug and play" different technological options into their farming systems, the reality is that agricultural technologies always interact with socioeconomic and biophysical factors. Agricultural technology is fundamentally different from software that can easily scale across similar hardware systems. As I discussed in the last chapter, we could apply a product development lens to agricultural development. But the "products" will mostly be context specific, and this is not a bad thing. This will allow organizations to better target small and marginal farmers if that is their goal.

A second lesson is that no other agricultural research program has achieved the "success" of widely adapted spring wheat, and even that success is qualified. Spring wheat is a particularly adaptable crop thanks to its genetics, photoperiod insensitivity, and responsiveness to fertilizers. Wheat farmers who most benefited from the Green Revolution shared socioeconomic and biophysical factors such as access to irrigation, fertilizers, and mechanization. To reach farmers that fall outside these bounds, a research program must target these more specific groups and not assume that benefits will spill over to noncommercial farmers.

The final lesson is that many champions of agricultural research for development do not appreciate the institutional innovations that facilitated the Green Revolution. Land reform, credit reform, and investment in fertilizers all allowed some farmers to quickly adopt Green Revolution technologies and to make a profit. These preconditions are not available to many farmers in less developed countries. The cost of agricultural inputs and land in some places have risen above the revenue from selling crops, leading some farmers to take on debt.

The institutions of agricultural research have also changed since the Green Revolution. Scientists like Borlaug had access to large budgets and long timelines. Today, donors demand results on fixed budgets and much shorter timelines. Furthermore, they expect technologies and practices to quickly scale over diverse agricultural systems. While this strategy makes sense for economic investments, it is not best the way to improve smallholder livelihoods. Last century's wide adaptation has become today's "impact at scale"—both efforts focus primarily on plant breeding and spreading technologies across vast scales. And both efforts will fail without recognizing the biophysical limits of wide adaptation and the socioeconomic context of agricultural innovation.

CIMMYT AND WIDE ADAPTATION FROM THE 1980S TO THE PRESENT

I now highlight the debates around wide adaptation from the 1980s forward, focusing on CIMMYT's continued adherence to the doctrine of wide adaptation and criticisms of this approach. My analysis is based primarily on published documents from CIMMYT or CIMMYT-affiliated researchers, as well as other published literature on the topics of adaptation and selection environments. While in the 1960s Borlaug's concept of wide adaptation was accepted in the plant breeding scientific community, it was never without controversy. This controversy continues today, though it often plays out behind closed doors and then occasionally bubbles to the surface as a research paper.

It is worth noting that debates over wide adaptation are rare within public and private plant breeding in the United States. Because the US agricultural research system is very robust, varieties are *always* specifically adapted to local conditions to maximize yield (although some varieties may have wide adaptation, that is not the goal). The point of wide adaptation is to produce varieties that can be more easily adapted to a range of conditions in areas with less scientific infrastructure. They may not be a perfect fit, but they can have higher yields than local varieties. Therefore, wide adaptation is mostly discussed in an agricultural research for development context.

Generally, scientists who support a more agroecological, biodiverse, and locally relevant type of agricultural research are skeptical of wide adaptation. Higher-ranking scientists at CIMMYT and in Indian wheat research tend to support wide adaptation. As I saw from my interviews, however, there is considerable variation among Indian wheat researchers on this topic, and we can expect the same from CIMMYT researchers. Apparently, younger CIMMYT scientists especially are more questioning of the mainstream approach.

At CIMMYT, Borlaug's continued influence is also clear. Borlaug worked there until 1979 but remained a consultant after he left. Subsequent directors of CIMMYT's global wheat program have been strong defenders of wide adaptation (including R. Glenn Anderson, R. A. "Tony" Fischer, Sanjaya Rajaram, and Hans-Joachim Braun). While CIMMYT has changed since the 1960s—for example, it no longer releases finished varieties to national programs—concepts such as mega-environments have deep roots in the philosophy and culture of wide adaptation. The result is that scientists and administrators at these organizations have a bias toward centralized systems and a belief in the scale-neutral technology. Without addressing these biases, these organizations will continue to have major blind spots in their programs aimed at poverty reduction and environmental sustainability.

Several studies in the 1980s and 1990s were critical of wide adaptation and its role in plant breeding. Some scientists argued that plant breeders should exploit genotype by environment interactions by using specifically adapted genotypes. Norman W. Simmonds, a plant breeder from the United Kingdom, was one prominent advocate of specific adaptation. He argued that the yield advances of the Green Revolution were due to improvements in environment rather than genetics and suggested "a deliberate policy of selecting and testing in well-chosen low-input environments" to counter this effect.[133]

While Simmonds made a mild suggestion toward specific adaptation, Salvatore Ceccarelli launched an attack on wide adaptation while working at ICARDA. Ceccarelli pioneered new techniques of decentralized plant breeding for specific adaptation of crops. In 1989 he published an article titled "Wide Adaptation: How Wide?" in which he criticized the concept of wide adaptation. Ceccarelli's CGIAR affiliation and his article created a problem for CIMMYT. Ceccarelli argued that breeding programs such as CIMMYT's had focused on higher-input environments to the detriment of low-input or marginal environments. He argued that while widely adapted varieties do exist, they exist only "*within* a given range of environments."[134] He found that widely adapted varieties frequently express what are called crossover interactions. Consider the Finlay-Wilkinson model, then imagine if the model is applied to two varieties. The analysis of one variety shows that it starts low on the y-axis but its slope rises quickly. The analysis of the other variety shows it starts higher on the y-axis, but has a flatter slope. The two lines would intersect at some point, demonstrating the "crossover" point, or the environment in which the two varieties have the same yield. This intersection means that one of the varieties has a lower yield in less-ideal environments, but higher yields in ideal environments. In practice, this is known to happen when improved varieties have higher average yields than local varieties, but those local varieties perform better in marginal environments. CIMMYT, at the time and subsequently, has claimed that their varieties do not cross over in marginal environments, and that on average they are universally superior to local varieties.

Articles by Simmonds and by Ceccarelli, W. Erskine, John Hamblin, and Stefania Grando both addressed the international wide adaptation strategy used by CIMMYT and others.[135] Simmonds wrote that in the Green Revolution, "selection has inevitably, but unconsciously, been for the high yielding, high-*b* [environmental responsive] variety" rather than for yield stability.[136] Ceccarelli and his coauthors argued that "international breeding programs must adopt a positive interpretation of genotype by environment interaction if they are to address the need of small, resource-poor, subsistence farmers, who have been by-passed by the Green Revolution," and used ICARDA's barley and lentil programs as an example of successful crop adaptations for low-input environments.[137] A different set of authors wrote that varieties developed by the international centers were "widely adapted to a range of environments" but "specifically adapted to none."[138]

Ceccarelli argued that the more marginal an environment, the more important it is to have varieties specifically adapted to that environment.[139] He has maintained that for marginal and stressed environments, selection of materials should occur in that target environment, or at least in contrasting environments, and not only in higher-input environments.[140] Simmonds agreed that selection for marginal environments must occur in situ, and argued further that breeding in contrasting environments (shuttle breeding) is ineffective and rarely systematically practiced.[141] A review by Peter Tigerstedt in 1994 supported these views, stating that "generally at the species margin, adaptation is of first order importance, yield comes second."[142] Ceccarelli found that despite numerous reviews on the topic of adaptation and stability showing that crossover interactions are a common occurrence, "specific adaptation is not a very popular concept. Presumably because it does not bring the same prestige to the breeder as the concept of wide adaptation and is not in the interest of the seed industry."[143]

CIMMYT's approach of shuttle breeding and multilocation testing for wide adaptation remained their primary method for wheat breeding, despite these criticisms. Other major CGIAR research centers also followed this approach.[144] CIMMYT's 1992 retrospective justified their focus on breeding for ideal environments because of the "spillover effect, especially noticeable in wheat, whereby the yield gains achieved under optimum moisture and fertility may also be reflected under less favorable conditions."[145] Acknowledging that new varieties may not perform as well in marginal conditions, they still asserted that "seldom does the improved genotype perform worse than the local variety."[146] This report also emphasized a commitment to wide adaptation, though it noted that they had achieved less success in widely adapting maize due to the diverse environments in which it is grown.

A 1996 conference on wheat improvement brought together Sanjaya Rajaram of CIMMYT and ICARDA's Ceccarelli for a paper in the proceedings of *Wheat: Prospects for Global Improvement* titled "International Collaboration in Cereal Breeding." These scientists discussed the fact that most breeders focus on yield alone, but this is not always the most important factor for farmers. For example, avoiding lodging is important in productive environments, but not always in dry environments, where farmers might want taller plants so that they are easier to harvest by hand. Rajaram and Ceccarelli said that international centers should stop providing countries with finished varieties, and instead give them earlier varieties to test in different environments within the country. They gave the example of ICARDA, which started doing this with barley in four countries in North Africa in 1991.

But a set of papers in the mid-1990s asserted CIMMYT's philosophical commitment to breeding for high yields and wide adaptation and their methodological commitment to shuttle breeding.[147] A 1994 paper by D. Steven Calhoun et al. of CIMMYT supported shuttle breeding between optimal and rainfed environments. They stated that their data "support the hypothesis that the highest yielding genotypes under favorable conditions will also be among the highest yielding under less favorable conditions."[148] CIMMYT has over time justified its continued focus on breeding under optimal conditions and shuttle breeding paired with multilocation testing under varied environments. CIMMYT scientists have argued that this is the best way to ensure higher heritability of genetic traits.[149] This view has not been universally accepted, however.

Starting in the 1980s, breeding for drought tolerance became a major international discussion in agricultural science, as scientists realized that most of the benefits from the Green Revolution occurred in irrigated areas. The idea that crops bred specifically for marginal conditions may have better adoption in those environments entered the discourse once again. The wheat breeder E. A. Hurd, from Canada, published several papers on drought resistance in self-pollinating crops in the late 1960s and 1970s. Hurd evaluated both the breeding approach and morphological and physiological characteristics that led to drought resistance.[150] Hurd's publications added a "discussion of philosophy" on "breeding for specific adaptation to target environments."[151]

Tony Fischer, on the other hand, promoted wide adaptation as a drought resistance strategy.[152] These contrasting but mainstream views led to a debate over "the dilemma facing the breeding when selecting for yield and environmental stress resistances" and "whether to breed for narrow adaptation to a specific environment . . . or to breed for wide adaptation."[153] Proponents of specific adaptation argued that to select for drought tolerance, varieties should be exposed to water stress during early breeding experiments (using what is called early generation or segregating material). Wide adaptation supporters argued that breeding in ideal environments and multilocation and multiyear testing of finished (stable) varieties resulted in varieties that were higher yielding and yield stable in both favorable and water-stressed conditions.

A highly cited paper by A. A. Rosielle and Hamblin continued the conversation on plant breeding in stressed versus nonstressed environments. Their study found that selecting for stress tolerance in stressed environments led to reduced yields in more favorable environments.[154] In other words, there was a tradeoff between stress tolerance and mean yield. Ceccarelli's 1987 article on barley breeding at ICARDA also confirmed the tradeoff between breeding for stress and breeding for yield, and suggested that breeding for dry environments should occur in the target environment.[155] Ceccarelli's study was based on a "parallel selection" method initiated by ICARDA's J. P. Srivastava et al. in a 1978 paper on durum wheat. In this study, the researchers used multilocation testing of still-segregating (early) generations.[156] They found that "selection under low fertility dry farming conditions is more successful in identifying drought tolerant genotypes than selections in the fertile high moisture environment."[157] Srivastava's statement on selection under suboptimal conditions contradicted what CIMMYT researchers argued.

CIMMYT addressed drought tolerance and their breeding philosophy in a 1984 wheat report. The authors noted that 37 percent of the wheat-producing area in developing countries is constrained by moisture (known as semiarid and arid regions), and that drought resistance is therefore the most important trait for new wheat varieties. The report stated that although drought resistance is often location-specific, varieties with "high yield, high stability, and wide adaptation" can be combined with drought resistance.[158] They also claimed these varieties could be developed ex situ, meaning bred in a different location than where they will be grown. A 1986 workshop report whose authors included CIMMYT scientists also noted that breeding for drought resistance faced challenges under the policies of national research systems. The authors recommended that national research systems allow more flexibility for varietal testing and release.[159]

In the late 1980s CIMMYT classified wheat-growing regions into twelve Mega-Environments.[160] These environments were based on agroclimatic conditions, primary biotic and abiotic stresses, and production practices (for example, rainfed versus irrigated). The idea was that wheat breeding would be targeted to each Mega-Environment, although critics have argued that most breeding is still targeted to optimal environments. CIMMYT also operates several international nurseries in different wheat environments. Research in these nurseries has led to findings such as that targeted breeding for higher temperatures can produce varieties better adapted to those conditions.[161]

The legacy of wide adaptation is still clear in both CIMMYT's leadership and its plant breeding philosophy. Hans-Joachim Braun, who supported wide adaptation and selection in optimal environments, led CIMMYT's global wheat program from 2004 to 2020 and the CGIAR Research Program on Wheat (WHEAT) from 2014 to 2020. Braun and other CIMMYT scientists have described the potential for "wide adaptation to buffer temporal climatic variability in wheat."[162] CIMMYT's approach to climate adaptation focuses on breeding for abiotic stresses such as drought tolerance, although there is also a focus on conservation agriculture. It is obvious that varietal improvement and wide adaptation will play a central role in CIMMYT's climate adaptation research program.

The CGIAR Research Program on Wheat was formed in 2012. WHEAT is led by CIMMYT with ICARDA as a collaborator. The program's goals are "reducing rural poverty and improving food and nutrition security in developing regions and enhancing sustainable management of natural resources."[163] But the continued focus on varietal improvement is clear. Over 60 percent of WHEAT researchers surveyed felt that "influencing food security through global germplasm development" is the "primary impact pathway."[164] Not surprisingly, most of WHEAT's research funding goes toward developing "productive wheat varieties."[165] The majority of the surveyed researchers also felt that most emphasis should be on "strategic research to promote international public goods" rather than "adaptive research" or "local research to produce feedback and synthesis."[166] Most of WHEAT's first strategic initiatives were focused on genetics.[167] This focus on upstream technologies without consideration of the context of adoption and production is certainly the result of decades of reliance on wide adaptation to justify centralized and decontextualized research.

Over the past few years CIMMYT and the CGIAR have put more stock in what one insider called the "molecular-genomic-speed-breeding camp." CIMMYT recently appointed Alison Bentley as director of CIMMYT's Global Wheat Program and WHEAT after Braun's retirement. Bentley's background is in wheat genetics, wheat genetic resources, and wheat prebreeding. Historically, directors of CIMMYT's Global Wheat Program have come from international research backgrounds. This hire signals one of many shifts in the CGIAR and CIMMYT away from engaging in local contexts and toward centralized, molecular approaches to agricultural research.

CIMMYT's responses to criticisms of the Green Revolution were in line with their parent organization, the CGIAR. By the late 1970s the CGIAR did recognize the failure of Green Revolution crops to reach farmers in marginal environments. The CGIAR implemented a program in farming systems research to correct this.[168] But alternatives to top-down breeding failed to thrive in most CGIAR research centers, and they have remained focused on centralized, more basic research rather than adaptive research.[169] As public funding for plant breeding and agricultural research have declined over time, plant breeders have become more entrenched in their position. They believe that focusing on other agricultural development activities will distract donors from the real solutions, which they believe are, of course, plant breeding. But the ideological commitment and bias toward to centralized plant breeding over adaptive research deserves serious reconsideration. The centralized approach assumes that countries have both the research and extension capacity to adapt technologies to their conditions. While this was true in Mexico, Pakistan, and India, it is not true of many developing countries, especially in sub-Saharan Africa.

CIMMYT has over the decades produced several wheat varieties that are high yielding and widely adapted, especially the Veery lines developed by Rajaram in the 1980s. These varieties have been popular in heat-stressed areas, including in India as the variety PBW343.[170] However, it is not clear whether the kind of quantum leap in yields and adaptability of Veery and the Green Revolution varieties is possible again.

But research on the relationship between varietal adoption and livelihood improvements is lacking. Vijesh Krishna et al. found that "most adoption studies have oversimplified and decontextualized technological change in wheat," partly because of donors demanding results on short time scales.[171] And as the authors point out, most of these studies view adoption as a binary status (adopted/nonadopted). It is important to remember that varieties usually require fertilizer and irrigation to produce higher yields and offset the seed cost. These inputs may not be available to smallholder farmers, and there are other mediating socio-economic factors such as market access.

To its credit, in recent years CIMMYT has hired some staff who have engaged with more critical conversation around top-down research and technology transfer. Hopefully, a better understanding of the links between technical change in agriculture and social and economic impacts will shape research programs for agricultural development. Because ultimately, improving yields of cereal crops may not be the most efficient approach to reducing poverty, food insecurity, and malnutrition.

THE LEGACY OF WIDE ADAPTATION IN INTERNATIONAL AGRICULTURAL DEVELOPMENT

Before concluding this book, I want to spend some time on what I call the "uniqueness of wheat." While much of today's international agricultural research system is built on the model of Green Revolution wheat, there are two aspects of wheat's "success" that are unique. These factors are not easy to replicate in other crops and farming systems. The first is wheat's unique genetic properties. Even before Borlaug's wheat program, wheat was known to have wide adaptability. This is due in part to wheat's hexaploid chromosome, which makes wheat's genome more complex than rice or maize.[1] Hexaploidy makes wheat harder to hybridize or modify through biotechnology, but it also makes wheat more flexible in different environments. True widely adapted wheat must have photoperiod insensitivity, which is also genetic. Some other crops such as rice and sorghum can be photoperiod insensitive, but scientists have not been able to exploit this trait in tropical maize. The second factor is that wheat-growing regions around the world tend to be agronomically similar, especially those under commercial cultivation. Other crops are generally grown under more diverse conditions. Wheat became the model for a global crop program, but we should view it as an outlier rather than an ideotype.

The uniqueness of wheat was clear in post–Green Revolution India. Semidwarf wheats quickly spread, while maize failed to launch and rice required local adaptation. In fact, Green Revolution–era spring wheats are said to have spread more quickly than any other innovation in the history of agriculture.[2] As of 1978 Indian experts pointed out that whereas adoption of "wheat spread at an extraordinarily high rate, those of paddy and other cereals have been trailing badly behind."[3] Kapil Subramanian showed that wheat was the only crop with a higher annual yield growth rate during the Green Revolution in India.[4] B. Sivaraman asked why "nobody paused to enquire as to why the scientists in nearly 2000 demonstrations, supervised closely by them on good farmers' lands, could not repeat their claims uniformly in the case of rice, maize, *jowar* [sorghum] and *bajra* [pearl millet]."[5] Wheat yields in India are actually higher than in the United States, yet India still struggles with malnutrition; 36 percent of children younger than five are underweight.[6] And despite the abundance of wheat, almost 20 percent of India's budget is spent on various subsidies and programs to lower the price of food.[7]

The agronomic advantages of wheat are also clear when comparing regions and crops. The areas in India and Mexico to first adopt the semidwarf wheats were the irrigated, commercial regions. In Colombia, however, wheat farmers tend to be poor and farm in the highlands. This is where the wide adaptation of wheat falls apart. The Rockefeller Foundation had been working in Colombia throughout the 1950s and 1960s, and some of the Green Revolution varieties were bred from Colombian wheats. But Colombian farmers were reluctant to adopt the new varieties. In recent years, wheat production in Colombia has been marginal. Less than 1 percent of Colombia's consumed wheat is produced domestically and the rest is imported.[8] Colombia's wheat yields are less than half of Mexico's, yet for maize and rice, the yields are similar.[9] This example shows that while wheat has a great deal of geographic flexibility, it is still limited by both climatic and economic parameters. In the case of both wheat's genetics and wheat-growing environments, it is difficult to apply these to other farming systems. Understanding these factors, we can better appreciate the context specificity of technical change in agriculture.

In this book I have put plant breeding at the center of the Green Revolution narrative. But you could center the narrative on either irrigation or fertilizer to explain the increase in cereal crop production during the Green Revolution.[10] There is a reason why histories of the Green Revolution tend to focus on plant breeding and crop varieties over these factors. The founders of the Green Revolution emphasized plant breeding as a scale-neutral science. This allowed them to deflect criticisms that they were not focused on small farmers or inequality. Similarly, scientists used wide adaptation to justify their focus on irrigated, well-fertilized farms. They argued that widely adapted crops also had high yields under rainfed and low-fertility conditions. In both cases, scientists use flawed concepts to justify how centralized research could benefit many.

Norman Borlaug was a key figure in these arguments, not just because of the varieties he developed but because of his activities shaping international politics and agricultural science. In other words, he leaves a legacy beyond wheat. One reason wide adaptation became so ingrained in international wheat research is Borlaug's strong personality. Borlaug possessed a "missionary zeal" that inspired his "wheat apostles," the scientists from the Middle East and Asia he trained in Mexico. Others have characterized Borlaug as a "brand hero" of the Green Revolution who continued to promote the Green Revolution approach until his death in 2009.[11] He also ridiculed those who questioned his methods or ideas. According to some scholars, "Borlaug used his growing recognition within the agriculture, science and policy communities to close down consideration of other views and methods. What could 'butterfly chasing' academics, environmentalists or bureaucrats possibly know about the realities of hunger fighting"?[12] Borlaug stated in 1986, "These critics were utopian intellectuals speaking from privileged positions in ivory towers who had never personally been hungry or ever lived and worked with people living in abject poverty."[13]

It is clear, however, that Borlaug was aware of critiques of his research method and chose to ignore them. As Harwood has pointed out, we can criticize the "occupational myopia" of agricultural scientists without calling it a moral failing.[14] Most experts share this narrow focus due to their specialized training. Charles C. Mann noted that in graduate school, Borlaug took classes only in botany and plant pathology.[15] He did not study any agronomy, soil science, hydrology, economics, or history. Borlaug was not a strict technologist, though. He was deeply involved in lobbying policymakers and scientists to favor a certain mode of agricultural development. "The psychology," he recalled in his 1967 oral history, "is as important as the technological change that you bring in (and everybody thinks that this [technology] is all that's needed)."[16] Borlaug was referring to the sense of competition he created between Indian and Pakistani agricultural scientists, but his advocacy went beyond that. Unfortunately, this lesson of the Green Revolution never became part of the narrative.

Borlaug's commitment to the Green Revolution research methods limited alternative ones, such as farming systems research, from achieving mainstream success. In his later years, Borlaug remained a strong advocate for technology-centric agricultural development. James Sumberg, Dennis Keeney, and Benedict Dempsey hypothesized that "without Borlaug (or someone playing a similar role), interest in the Green Revolution approach may have been more difficult to sustain, and more space might have opened up for debate about alternative approaches and methods."[17] With Borlaug's strong involvement in CIMMYT and transferring those lessons to other CGIAR centers, however, the Green Revolution methods and the goal of increased food production became the international standard for agricultural research.

Borlaug passed away in 2009 at ninety-five years old, and his story has had a strong emotional impact on many people in the world, including me. Some think that because of Borlaug's legacy, we should not criticize him. I disagree. I believe that no scientist should be shielded from thoughtful, well-reasoned criticism, and that it is their professional duty to consider these critiques, acknowledge nuances, and demonstrate humility. Borlaug showed little evidence of this.[18] At some point, we might admit that our heroes never live up to our hopes. That we humans are all flawed, and Nobel Prize winners are no exception. They might even have moral failings. But perhaps it is less important to retrospectively judge Borlaug than to change our view of him from infallible hero to a hardworking, passionate scientist who nevertheless had human flaws that influenced his perspective and work. This alone might open agricultural development to new pathways and ways of thinking.

In Borlaug's wake came a new brand hero for agricultural development—not a scientist, but a software engineer and philanthropist. Bill Gates and his foundation, the Bill and Melinda Gates Foundation, have supported agricultural development in Africa since 2006. Gates has specifically focused on small farmers and has stated, "Melinda and I believe that helping the poorest small-holder farmers grow more crops and get them to market is the world's single most powerful lever for reducing hunger and poverty."[19] In this speech Gates reflected on making the Green Revolution more sustainable and targeted at small farmers. He mentioned working with local farmer groups. But the most visible agricultural efforts by the Gates Foundation have focused on technologies, such as funding the development of drought-tolerant maize in Africa and supporting the Alliance for Science, which promotes public understanding of agricultural biotechnology. While recognizing some negative consequences of the Green Revolution, the Gates Foundation's agricultural work follows the same basic model of the original Green Revolution: centralized research and technology development.

As Harwood so clearly explains, promotors of an African Green Revolution have "remarkably little to say about how this revolution will manage to avoid the adverse consequences of the original GR [Green Revolution]. If they believe that 'this time it will be different,' they have not made clear why that should be the case. Their silence raises the disturbing possibility that proponents may once again be burying their heads in the sand, ignoring what happened the last time around."[20] The Green Revolution narrative persists.

The focus on technology-centric solutions is not just in the Gates Foundation. It is also in the US Agency for International Development and the CGIAR generally.[21] CIMMYT recently launched the program Accelerating Genetic Gains in Maize and Wheat for Improved Livelihoods, which claims to be "better, faster, equitable, sustainable."[22] This effort builds on a previous project at Cornell University called Delivering Genetic Gains in Wheat, and is funded by the Gates Foundation and US and UK government aid. These efforts are modeled on the Green Revolution using genomic research techniques but also claim to improve equity, sustainability, and smallholder livelihoods. A press release for the Genetic Gains project stated, "The project specifically focuses on supporting smallholder farmers in low- and middle-income countries. The international team uses innovative methods—such as rapid cycling and molecular breeding approaches—that improve breeding efficiency and precision to produce varieties that are climate-resilient, pest and disease resistant and highly nutritious, targeted to farmers' specific needs."[23] It is unclear how this program will actually determine the "specific needs" of farmers in diverse environments or target technology deliver to smallholders. What is clear is that some of the highest-profile agricultural development projects are still focused on genetic improvement of yield, with little attention to the surrounding agroclimatic and socioeconomic systems.

Smallholder farmers are presumably more vulnerable to climate change than commercial farmers. Public international agricultural research has focused on climate-tolerant varieties as the primary adaptation pathway for these farmers. There are other ways for farmers to adapt to climate change, such as crop and livestock diversification or pursing nonagricultural employment.[24] But the idea of climate-tolerant varieties fits neatly into the CGIAR's and Gates Foundation's paradigms of centralized research that can be scaled across environments. These programs aim to benefit smallholder farmers first, but this theory of change clearly lacks a social science lens. There are virtually no historical examples of centralized research models that benefit smallholder farmers before larger or commercial farmers. Along the same lines, Klara Fischer shows that "scale-neutral" technologies are becoming a popular concept again: "The term is now reappearing in the debate on the potential of new crop varieties in general, and genetically modified (GM) crops in particular, to benefit African smallholders."[25]

The Gates Foundation–funded Alliance for a Green Revolution in

Africa is focused on improved crop varieties, fertilizer use and management, and value chain integration.[26] The Alliance for a Green Revolution in Africa claims that improved technology, management, and market access will reduce rural poverty. Mary Ollenburger et al., however, showed that in Mali, "the potential gains from intensification of dryland agriculture—whether sustainable or not—are not competitive with off-farm options for most farming households."[27] Ollenburger's analysis concurs with David Harris and Alastair Orr's finding that increased yields seldom lift smallholder farmers out of poverty.[28] An evaluation of the Gates Foundations' seed system program found that in areas targeted by the Alliance for a Green Revolution in Africa for ten years, three regions showed no difference in the poverty levels of households and one showed a weakly significant improvement.[29] Even a CGIAR evaluation report remarked that "higher yields can eliminate poverty for those subsistence farmers who can become commercial farms; only a small share of rural households, though, seems capable of making the transition."[30]

In an excerpt from his recent book, Gates claims that the CGIAR is one of the world's best bets for reducing rural poverty and increasing rural resilience to climate change.[31] About 20 percent of his foundation's agricultural development budget has gone to CGIAR-affiliated programs.[32] Gates explains that his favorite pro-poor agricultural innovation is drought-tolerant maize, which his foundation has funded through CIMMYT for over a decade. CIMMYT scientists bred these varieties using the CIMMYT strategy of breeding for large agroecological zones and then relying on national programs to adapt to local conditions. Sally Brooks et al. called this effort "streamlining a technology supply 'pipeline' in order to achieve 'impact at scale.'"[33] Rachel Schurman notes that the Gates Foundation focuses on "having impact on a large scale," which "leads the foundation to privilege big, international organizations that can develop and manage megaprojects" (such as the CGIAR over local organizations) and manifests as "making sure that its grants can be 'scaled up.'"[34] Again, this is not just the Gates Foundation focused on scaling, but part of a broader trend in agricultural development where donors demand bigger results on shorter timescales.[35]

Drought-tolerant maize for Africa could in theory raise yields during dry seasons and improve farmers' resilience to climate variation. But there are several flaws in the project's approach, which replicates rather than improves the Green Revolution model. Social scientists have warned for over a decade that focusing on drought-tolerant maize will not improve farmers' resilience but rather lock them into a maize system.[36] This effort also overlooks that most African countries have poorly funded research and extension systems, which were key enabling factors in the Asian Green Revolution. Brooks et al. also warned the structure of this project sidelines participatory research and does not adequately acknowledge the agroclimatic diversity faced by smallholder farmers in Africa.[37]

Farmers in Africa have been slow to adopt drought-tolerant maize varieties.[38] There are also gendered differences in adoption, and larger smallholder farmers were more likely to adopt drought-tolerant maize in Zimbabwe.[39] Similarly, flood-tolerant rice varieties have had low levels of adoption in South Asia.[40] These low rates of adoption may not reflect the appropriateness of the technology but rather the structure of the surrounding socioeconomic systems. Smallholder farmers in Africa do not buy much seed from formal seed markets, where these varieties would be distributed. The Gates Foundation has funded seed system development, but focused on education of scientists and interventions in the private seed sector and not the informal seed system that most smallholders rely on.[41] There is also a bigger question about whether alternative climate adaptations could be more effective than promoting a single, resource-intensive crop.

These outcomes are not surprising to anyone familiar with the social science literature on the Green Revolution. Schurman reminds us that "*context* is crucial for shaping the consequences of agricultural development interventions."[42] The question remains, then, why do major development organizations continue to pursue the Green Revolution strategy and claim it will benefit smallholder farmers? Many of the preceding chapters have shown that the technology-driven strategy helps scientists maintain their power. But Gates and most of his staff are not scientists, so this is not a complete explanation. I propose that answer is related to "the apparently irresistible lure of a technical fix for all manner of socioeconomic problems."[43] Crop-based interventions are more compatible with the Gates Foundation's "attraction to technological fixes and the 'scalability' of its investments" than other social, economic, and political interventions.[44] These alternatives require much more political capital and longstanding local connections and are not easily scaled across contexts. There is a fundamental tension between the scalability of technologies and the diversity of smallholder farming conditions. But Green Revolution concepts like wide adaptation and scale neutrality promote the myth that technologies can easily scale across both biophysical and socioeconomic contexts. These concepts are central to current visions of extending the Green Revolution, but deserve a closer look.

Historically successful strategies for improving smallholder livelihoods have not been forgotten so much as they have been willfully ignored by the dominant agricultural research organizations.[45] There is also a valid question of whether cereal crop improvement can contribute to poverty reduction or nutritional security. Ollenburger et al. summarize the outlook for international agricultural research: "To improve rural livelihoods while also increasing the production of staple foods to feed a growing population, it is vital that researchers, policy makers, development practitioners and other stakeholders find ways in which their goals can intersect with farmers' priorities rather than simply imposing their own goals on rural communities." They continue, "If the goal is to improve smallholder livelihoods, agricultural interventions directly linked to food production must be accompanied by efforts to address the priorities rural people themselves identify. . . . If they do not take into account existing social and ecological conditions and respond to farmers' priorities, the intensification practices proposed by many agricultural development institutions may simply be solutions in search of a problem."[46]

Agricultural research must move beyond the narratives of the Green Revolution to understand that globalization has changed the relationship between poverty and agricultural production. We should remember that Borlaug's ideas were radical at the time, but he was given the freedom, budget, and timeline to pursue them. To improve smallholder livelihoods, we need to give them more than lip service—we need to strengthen local research systems, encourage context-appropriate innovations, and look beyond technological solutions. Seriously addressing rural poverty, inequity, and food security requires both learning from the past and an openness to exploring paths yet uncovered.

RESEARCH METHODS

HISTORICAL RESEARCH

This book relies on historical documents that I collected at various depositories in 2013 and 2014. I selected documents based on their relevance to wheat's agroclimatic adaptation, the institutional history of relevant organizations, and correspondence between relevant actors. My initial research was restricted to the period of 1950 to 1970, but at the Rockefeller Archive Center and from online archives I also collected documents from the 1970s and later.

INDIAN ARCHIVES

In India I drew materials from several agricultural research libraries at the Indian Agricultural Research Institute (IARI) in New Delhi, the Directorate of Wheat Research (DWR) in Karnal, Haryana (named ICAR–Indian Institute of Wheat and Barley Research at the time of publication), and Punjab Agricultural University in Ludhiana, Punjab. While these libraries did not host any archives, they contained both primary and secondary sources. The IARI library hosts a wealth of annual reports and conference proceedings from the middle of the century (and earlier) to the present. Sources included published and unpublished material including annual reports, national and international wheat conference proceedings, journal articles, and published books. The annual wheat research workers' workshop proceedings were collected from both the IARI library and the DWR library.

HISTORICAL WHEAT MULTILOCATION TRIAL DATA

I used reports of the All India Coordinated Wheat Improvement Program to collect both qualitative and quantitative data on the prevalence and performance of wheat varieties under different conditions. I collected data for every year available from 1965 to present, for the Advanced Varietal Trials (formerly Uniform Regional Trials) and the Northwestern Plain Zone and Northeastern Plain Zone.

ROCKEFELLER ARCHIVE CENTER

I spent three weeks at the Rockefeller Archive Center in April and May 2014. While there, I focused on three topics: the Rockefeller Foundation's international wheat programs in the 1960s (focused on the office in Mexico), its Indian agricultural program in the 1950s and 1960s, and its international wheat programs in the 1970s (focused on the Middle East). For the international wheat programs in the 1960s, I was interested in the progression from the Mexican Agricultural Program to the Inter-American Food Crop Improvement Program, then to the International Center for Corn and Wheat Improvement, to eventually CIMMYT. To these ends, I examined records from the Rockefeller Foundation's project files (Record Group 1.2, Series 300D and 323, and Record Group 1.3, Series 105), administration, program, and policy files (Record Group 3, Series 915 and 923), Mexico field office (Record Group 6.13, Series 1), officer diaries (Record Group 12), and oral histories (Record Group 13), as well as the Rockefeller Foundation Agricultural Science Program Annual Reports.

I also focused on the Rockefeller Foundation's involvement in Indian maize and wheat improvement. Toward this end, I used project files (Record Group 1.2, Series 464D, and Record Group 1.6, Series 464D) and the New Delhi field office records (Record Group 6.7). Finally, I explored the Rockefeller Foundation's expansion into wheat research in the Middle East in the 1970s, using again the project files (Record Group 1.3, Series 105) and some recently added archival material from the Ankara, Turkey, field office (Record Group 6, Series 19). I also used some newly released (at the time) material from the Ford Foundation Archives, colocated at the Rockefeller Archive Center, on the Arid Lands Agricultural Development (ALAD) and International Center for Agricultural Research in the Dry Areas (ICARDA) programs in the Middle East.

IOWA STATE UNIVERSITY LIBRARY SPECIAL COLLECTIONS AND UNIVERSITY OF MINNESOTA LIBRARY

Because of the importance of Norman E. Borlaug to this research, I drew from two collections dedicated to Borlaug's papers. The Minnesota archive was only accessed online; thus, my browsing was less thorough and restricted to Borlaug's correspondence in the 1960s and his 1967 oral history. I spent one week at the Iowa State University Library Special Collections, where I focused on Borlaug's correspondence with Charles F. Krull and Keith W. Finlay, as well as Borlaug's and other Rockefeller Foundation scientists' involvement in various Food and Agriculture Organization of the United Nations programs.

FIELDWORK IN NORTHERN INDIA

Much of this research was conducted in northern India from January to July 2013. I was financially supported by an National Security Education Program Boren Fellowship and I was appointed as a Research Fellow with Bioversity International's subregional office in New Delhi under the guidance of Dr. P. N. Mathur.

While in India I traveled to several agricultural research institutions involved in wheat research. These included: IARI Regional Research Station in Pusa, Bihar (established in 1905, it is the birthplace of wheat research in India, and the former center of IARI until 1934), the Directorate of Wheat Research in Karnal, Haryana (established at Karnal in 1990), Punjab Agricultural University (established in 1962), and the IARI headquarters in New Delhi (established in 1936). I stayed at each institution for about one week each (except for the IARI, New Delhi), and spent two to three days at each place conducting interviews.

SCIENTIST INTERVIEWS

From February to May 2013, I interviewed forty-seven agricultural scientists at four major agricultural research institutions in India. Thirty-two of these interviews were with practicing scientists involved in wheat improvement and extension programs. Main fields of the scientists interviewed included plant breeding, biotechnology, genetics, plant protection (pathology), quality, agronomy, and extension science. Participants were chosen through snowball sampling. At each field site I had a local scientist host who would introduce me to the other scientists based on the criteria I had laid out of scientists involved in the Wheat Improvement Programme and extension scientists. The scientists I interviewed ranged from junior-level plant breeders to senior-level project directors. Overall, my data may be skewed toward more senior scientists, since that was who I was often directed to talk to.

Scientist interviews generally lasted between fifteen and sixty minutes and followed a structured questionnaire. Agricultural scientists were given a specific set of questions, while extension scientists were asked a separate set of questions. Interviews were conducted and record-

TABLE A.I. Number of recorded interviews with wheat scientists, by field and institution

	IARI Delhi	IARI Pusa	DWR	PAU	Total
Plant breeding	2	1	4	3	10
Biotechnology	1	1	4	0	6
Plant protection	1	0	1	3	5
Wheat quality	1	0	1	1	3
Agronomy	0	1	2	0	3
Seed Production	0	1	0	0	1
Extension	0	2	0	2	4
Total	5	6	12	9	32

ed in English and transcribed by the author. A few interviews were not recorded based on the informant's preference. Institutional review board approval was obtained for all interview questions (including for farmers and administrators), and all interviews were confidential (no names were recorded, and no identifying information is used in this book).

When circumstances made the questionnaire format incompatible with the situation (such as time constraints, interviews that became off-topic, scientists whose research did not apply to the questionnaire), I would ask unstructured questions that were based on the main themes of the questionnaire. Despite this, most of the interviews were structured and most of the questions could be answered to a satisfactory degree.

Based on completeness of the structured interview questions, twenty-five interviews were analyzed and coded. Answers to structured interview questions were put into a spreadsheet and analyzed for keywords and themes. Themes included wheat breeding for location, breeding for agroclimatic conditions, research on stress tolerance, microclimatic factors, and the structure of Indian research. Key statements were transcribed and quoted to reflect these themes, including themes to represent opposing viewpoints.

RESEARCH ADMINISTRATOR AND RETIRED SCIENTIST INTERVIEWS

While in India I interviewed eight current and retired agricultural research administrators. Due to varying time allotments for interviews with administrators, these varied greatly in both questions and content. Administrator interviews were semistructured and centered on Indian agricultural science policy in general and specifically related to wheat improvement in north India. Two retired wheat breeders at the IARI in

New Delhi were also interviewed. These interviews followed a similar format as the administrator interviews. Transcripts from the interviews with research administrators and retired scientists were analyzed qualitatively. Administrators and retired scientists were identified through my local research networks. In the process of writing and revising this book, I also engaged with a few retired international scientists to clarify information.

NOTES

INTRODUCTION

Epilogue: N. I. Vavilov, *The Origin, Variation, Immunity and Breeding of Cultivated Plants*, trans. K. Starr Chester (1935; repr., Waltham: Chronica Botanica, 1951).

1. See Oasa, "International Rice Research Institute," for discussion of how wide adaptation failed as a model for international rice research.

2. Borlaug, "Wheat Breeding," 8.

3. Perkins, *Geopolitics and the Green Revolution*.

4. Fischer, "Why New Crop Technology Is Not Scale-Neutral."

5. Freebairn, "Did the Green Revolution," 276.

6. Rapley, *American Experience*. Justin Cremer, "Norman Borlaug Saved Millions of Lives, Would His Critics Prefer He Hadn't?" Cornell Alliance for Science, April 24, 2020, https://allianceforscience.cornell.edu/blog/2020/04/norman-borlaug-legacy-documentary/.

7. Cremer, "Norman Borlaug Saved Millions."

8. "Ending Hunger," 336.

9. Nelson, Coe, and Haussmann, "Farmer Research Networks," 2.

10. While the Green Revolution started with the Rockefeller Foundation's research in Mexico and South America, this is secondary to the primary narrative around the Green Revolution in Asia.

11. Subramanian, "Revisiting the Green Revolution."

12. Richa Kumar, "Putting Wheat in Its Place, or Why the Green Revolution

Wasn't Quite What It's Made out to Be," *Wire*, October 31, 2016, https://thewire
.in/agriculture/green-revolution-borlaug-food-security.

13. Anderson et al., *Science, Politics, and the Agricultural Revolution*; Cleaver, "Contradictions of the Green Revolution"; Frankel, *India's Green Revolution*; Griffin, *Political Economy of Agrarian Change*; Ladejinsky, "Green Revolution in Punjab."

14. Harwood, "Was the Green Revolution Intended."

15. Sen, *Green Revolution in India*; Sisodia, "Photoinsensitivity in Wheat Breeding."

16. Freebairn, "Did the Green Revolution."

17. Cullather, *Hungry World*; Pielke and Linnér, "From Green Revolution to Green Evolution."

18. Spielman and Pandya-Lorch, *Proven Successes in Agricultural Development*.

19. Cullather, *Hungry World*.

20. Pielke and Linnér, "From Green Revolution to Green Evolution." The newly established Food Corporation of India has played a major role in the redistribution of food and keeping domestic food prices low. Subramanian, "Revisiting the Green Revolution."

21. Davis, *Late Victorian Holocausts*; Sen, "Ingredients of Famine Analysis."

22. Patel, "Long Green Revolution," 24–25.

23. UNICEF, *Malnutrition Rates Remain Alarming*.

24. Das, "Green Revolution and Poverty."

25. Mujib Mashal, Emily Schmall, and Russell Goldman, "Why Are Farmers Protesting in India?" *New York Times*, January 27, 2021, https://www.nytimes
.com/2021/01/27/world/asia/india-farmer-protest.html.

26. Fischer and Hadju, "Does Raising Maize Yields"; Harris and Orr, "Is Rainfed Agriculture."

27. Samberg et al., "Subnational Distribution," 1.

28. Singh and Agrawal, "Improving Efficiency," 387.

29. Nelson, Coe, and Haussmann, "Farmer Research Networks."

30. Harwood, *Europe's Green Revolution and Its Successors*.

31. Orr, "Why Were So Many Social Scientists."

32. Dominic Glover and Nigel Poole wrote that to achieve food security and sustainable agriculture, "the countries of South Asia need more innovation. However, they will also benefit from intelligent reflection about what innovation means, the directions it should take, and its risks and downsides alongside its benefits." Glover and Poole, "Principles of Innovation," 63.

33. There is also an increasing focus on improving farmers' access to markets; however, this too is primarily technologically focused.

34. See the reasons laid out by historian Jonathan Harwood in "Another Green Revolution."

35. Baranski, "Wide Adaptation of Green Revolution Wheat"; Harwood, *Europe's Green Revolution and Its Successors*; Kumar, "'Modernization' and Agrarian Development"; Olsson, *Agrarian Crossings*; Pielke and Linnér, "From Green Revolution to Green Evolution"; Saha, "State Policy"; Schmalzer, *Red Revolution, Green Revolution*; Soto Laveaga, "Socialist Origins of the Green Revolution"; Subramanian, "Revisiting the Green Revolution."

36. Subramanian, "Revisiting the Green Revolution"; Stone, "Commentary."

37. Stone, "Commentary," 7.

38. Harwood, "Was the Green Revolution Intended," 314.

39. Harwood, "Could the Adverse Consequences."

40. Fuglie et al., *Harvesting Prosperity*, 4.

41. Harwood, "Another Green Revolution."

42. Cabral, Pandey, and Xu, "Epic Narratives"; Cullather, *Hungry World*; Patel, "Long Green Revolution"; Pielke and Linnér, "From Green Revolution to Green Evolution"; Sumberg, Keeney, and Dempsey, "Public Agronomy"; Thompson and Scoones, "Addressing the Dynamics of Agri-Food Systems."

43. Pielke and Linnér, "From Green Revolution to Green Evolution."

44. Pielke and Linnér, "From Green Revolution to Green Evolution," 14.

45. Sumberg, Keeney, and Dempsey, "Public Agronomy," 1589.

46. Prasad, "Constructing Alternative Socio-technical Worlds," 293.

47. CGIAR was previously an abbreviation for Consultative Group on International Agricultural Research but the organization is presently just referred to as the CGIAR. Centro Internacional de Mejoramiento de Maíz y Trigo is commonly translated as the International Maize and Wheat Improvement Center.

48. Harwood, *Europe's Green Revolution and Its Successors*.

49. Bill Gates, "2009 World Food Prize Symposium," prepared remarks, October 15, 2009, https://www.gatesfoundation.org/Media-Center/Speeches/2009/10/Bill-Gates-2009-World-Food-Prize-Symposium.

50. Harwood, *Europe's Green Revolution and Its Successors*; Ollenburger et al., "Are Farmers Searching."

51. Brooks et al., *Environmental Change and Maize Innovation*; Harwood, *Europe's Green Revolution and Its Successors*; Pielke and Linnér, "From Green Revolution to Green Evolution."

52. Leach, Scoones, and Stirling, *Dynamic Sustainabilities*, 5.

53. Pielke and Linnér, "From Green Revolution to Green Evolution," 23.

54. Hughes, "Technological Momentum."

55. This is most obvious in the energy sector.

56. McGuire, "Path-Dependency in Plant Breeding."

57. Dahlberg, "Ethical and Value Issues."

58. Adas, *Machines as the Measure of Men*, 412.

59. Latham, *Right Kind of Revolution*.

60. Westad, *Global Cold War*.

61. Adas, *Machines as the Measure of Men*; Cowen and Shenton, *Doctrines of Development*.

62. Perkins, *Geopolitics and the Green Revolution*.

63. Bashford, "Population, Geopolitics, and International Organizations"; Cullather, *Hungry World*; Ramsden, "Carving Up Population Science."

64. W. W. Rostow to Dean Rusk, October 28, 1964, US Department of State, Foreign Relations of the United States, 1964–1968: International Development and Economic Defense Policy; Commodities, vol. IX, doc. 17, retrieved from http://history.state.gov/historicaldocuments/frus1964-68v09/d17.

65. Perkins, *Geopolitics and the Green Revolution*.

66. Ahlberg, "'Machiavelli with a Heart.'"

67. Ahlberg, "'Machiavelli with a Heart.'"

68. Perkins, *Geopolitics and the Green Revolution*.

69. Westad, *Global Cold War*. The term "Third World" is based on the French prerevolutionary "third estate."

70. Fitzgerald, "Exporting American Agriculture."

71. Lewontin, "Green Revolution."

72. Cullather, *Hungry World*.

73. Perkins, *Geopolitics and the Green Revolution*.

74. Jennings, *Foundations of International Agricultural Research*.

75. Escuela Nacional de Agricultura de Chapingo later became Estatuto de la Universidad Autónoma Chapingo (Chapingo Autonomous University).

76. Fitzgerald, "Exporting American Agriculture"; Perkins, *Geopolitics and the Green Revolution*.

77. Perkins, *Geopolitics and the Green Revolution*.

78. Curry, "From Working Collections"; Kloppenburg, *First the Seed*.

79. Bashford, "Population, Geopolitics, and International Organizations."

80. Excerpt from minutes of meeting of IHD Commission on Review, May 19, 1950, folder 20, box 3, series 915, record group [RG] 3, Administration, Program, and Policy, FA112, Rockefeller Foundation Records, Rockefeller Archive Center, Tarrytown, NY. Emphasis in the original.

81. Cullather, *Hungry World*; C. B. Fahs to Dean Rusk, January 12, 1954, folder 21, box 3, series 915, RG 3, Administration, Program, and Policy, FA112, Rockefeller Foundation Records.

82. J. George Harrar, "Agriculture and the Rockefeller Foundation," 1951, folder 20, box 3, series 915, RG 3, Administration, Program, and Policy, FA112, Rockefeller Foundation Records.

83. Stakman, Bradfield, and Mangelsdorf, *Campaigns against Hunger*.

84. Jennings, *Foundations of International Agricultural Research*.

85. Stakman, Bradfield, and Mangelsdorf, *Campaigns against Hunger.*

86. Oral history of Elvin Charles Stakman, 1966–67, folder 2, box 24, RG 13, Oral Histories, FA119, Rockefeller Foundation Records.

87. Oficina de Estudios Especiales, *Fourth Annual Progress Report.*

88. Colombian Agricultural Program, *Director's Annual Report: May 1, 1955–April 30, 1956.*

89. Colombian Agricultural Program, *Director's Annual Report: May 1, 1956–April 30, 1957,* 35.

90. Rockefeller Foundation, *Annual Report* (1959), 25.

91. Harrar to N. E. Borlaug, 1958, folder 110, box 14, series 300D, RG 1.2, Projects, FA387, Rockefeller Foundation Records.

92. Harrar to José Vallega, October 21, 1959, folder 110, box 14, series 300D, RG 1.2, Projects, FA387, Rockefeller Foundation Records.

93. Jennings, *Foundations of International Agricultural Research.*

94. Press release, November 29, 1963, folder 174, box 25, series 323, RG 1.2, Projects, FA387, Rockefeller Foundation Records.

95. Julian Rodriguez Adame, speech, October 25, 1963, folder 174, box 25, series 323, RG 1.2, Projects, FA387, Rockefeller Foundation Records.

96. International Center for Corn and Wheat Improvement, Report, 1963, folder 174, box 25, series 323, RG 1.2, Projects, FA 387, Rockefeller Foundation Records.

97. International Center for Corn and Wheat Improvement, report, 1963, folder 174, box 25, Series 323, RG 1.2, Projects, FA 387, Rockefeller Foundation Records.

98. L. M. Roberts, "The Desirability of Strengthening the International Center for Corn and Wheat Improvement and its Global Activities," 1965, folder 176, box 25, series 323, RG 1.2, Projects, FA 387, Rockefeller Foundation Records.

99. Jennings, *Foundations of International Agricultural Research.*

CHAPTER I: NARRATIVES AROUND WIDE ADAPTATION IN INTERNATIONAL WHEAT RESEARCH, 1960–1970

1. Annicchiarico, *Genotype × Environment Interactions.*

2. Pawley, *Nature of the Future.*

3. US Patent Office, *Report of the Commissioner of Patents,* 387. Agroclimate refers to the climatic and environmental characteristics of a location. An agroclimatic range or zone is an area that supports the cultivation of a certain crop, crop variety, or crop system.

4. Wide adaptation was pursued as a plant breeding strategy in Germany starting in the 1890s. Plant breeders aimed to breed "universal varieties" for a few decades until this strategy was discarded in favor of localized breeding. Harwood, *Europe's Green Revolution and Its Successors.*

5. Annicchiarico, *Genotype × Environment Interactions.*

6. Oficina de Estudios Especiales, *Fourth Annual Progress Report*, 10.

7. Perkins, *Geopolitics and the Green Revolution*, 12.

8. Smith, "Governing Rice."

9. Fitzgerald, "Exporting American Agriculture," 468–69.

10. A nursery, in this context, is a test of the same sample of varieties under different conditions and geographies. The "nursery" is the collection of seed samples, not the actual location in which they are grown. Barnes, *Transnational Flows of Expertise*.

11. Rusts are fungal diseases that can reduce the yield of crops. Different rusts affect different parts of the plant.

12. Oficina de Estudios Especiales, *Fourth Annual Progress Report*.

13. Elvin Stakman, memorandum to Dean Rusk regarding questions in attached letter, December 22, 1953, folder 21, box 3, series 915, record group [RG] 3, Administration, Program, and Policy, FA112, Rockefeller Foundation Records, Rockefeller Archive Center, Tarrytown, NY.

14. Stakman, memorandum to Rusk, December 22, 1953, 11.

15. Oral history of J. George Harrar, 1961–62, p. 38, folder 5, box 17, RG 13, Oral Histories, FA119, Rockefeller Foundation Records.

16. Loegering and Borlaug, *Contribution of the International Spring Wheat*.

17. Notes on the Consultants' Meeting—July 26 and 27, 1960, Mexico City, 1960, folder 11, box 2, series 923, RG 3, Administration, Program, and Policy, FA112, Rockefeller Foundation Records.

18. Norman E. Borlaug, "Report of South American Trip, Nov 11–Dec 18, 1959," folder 110, box 14, series 300D, RG 1.2, Projects, FA387, Rockefeller Foundation Records.

19. Borlaug, "International Wheat Research Project for 1960," 1959, folder 340, box 30, series 1, RG 6.13, Field Offices, Mexico, FA398, Rockefeller Foundation Records.

20. Borlaug to Robert W. Romig, June 24, 1960, MS 467, box 21, folder 64, Norman E. Borlaug Papers, Special Collections Department, Iowa State University Library, Ames, IA.

21. Borlaug, Ortega, and García, *Preliminary Report*.

22. Borlaug to A. H. Boerma, November 3, 1970, FAO Corporate Document Repository, http://www.fao.org/docrep/x5591e/x5591e0b.htm.

23. Byerlee, *Birth of CIMMYT*.

24. Byerlee, *Birth of CIMMYT*.

25. Notes on the Consultants' Meeting—July 26 and 27, 1960.

26. Borlaug to Romig, June 24, 1960.

27. Rockefeller Foundation, *Annual Report, 1959–60*, 254.

28. Borlaug, Ortega, and García, *Preliminary Report*.

29. Krull et al., *Results of the First International Spring Wheat Yield Nursery*.

30. Rockefeller Foundation, *Annual Report, 1963–4.*

31. Borlaug to Albert Moseman, January 18, 1963, coll. 01014, box 3, folder 3, Norman E. Borlaug Papers, 1930–2006, University of Minnesota Libraries, Minneapolis, MN.

32. Rockefeller Foundation, *Annual Report, 1964–5,* 214.

33. Rockefeller Foundation, *Annual Report, 1959–60,* 255.

34. Borlaug, "Wheat Breeding and Its Impact." Mentana had been developed by the Italian scientist Nazareno Strampelli for its rust resistance and earliness, incorporating the important trait of photoperiod insensitivity. Salvi, Porfiri, and Ceccarelli, "Nazareno Strampelli."

35. See Charles C. Mann's *The Wizard and the Prophet* for a detailed account of the painstaking breeding work Borlaug engaged in and more detail on the shuttle breeding method.

36. See Cullather, *Hungry World.* Borlaug's breeding method was not called shuttle breeding until the 1970s, when CIMMYT director Haldore Hanson "suggested that it be called 'shuttle breeding,' after U.S. Secretary of State Henry Kissinger's 'shuttle diplomacy' in the Middle East." CIMMYT, *Enduring Designs for Change,* 14.

37. Oral history of Norman Borlaug, 1967, p. 188, coll. 01014, box 1, folder 5, Borlaug Papers, University of Minnesota Libraries.

38. Hesser, *Man Who Fed the World.*

39. Borlaug, Ortega, and García, *Preliminary Report,* 7.

40. Greulach, "Photoperiodism."

41. Reynolds, Pask, and Mullan, *Physiological Breeding I.*

42. Borlaug, "Green Revolution," 115.

43. Borlaug, "Green Revolution," 115.

44. Rockefeller Foundation, *Annual Report, 1963–4,* 228, 229.

45. Cullather, *Hungry World.*

46. Borlaug and John Gibler, "Progress in Wheat Research in Argentina in 1964," January 12, 1965, p. 1, folder 181, box 26, series 323, RG 1.2, Projects, FA387, Rockefeller Foundation Records.

47. Borlaug, "Diary—NEB—Trip made to Argentina, June 27–July 15, 1962," folder 111, box 14, series 300D, RG 1.2, Projects, FA387, Rockefeller Foundation Records.

48. Borlaug to M. V. Rao and J. P. Tandon, January 6, 1987, MS 467, box 21, folder 14, Borlaug Papers, Iowa State University Library.

49. Borlaug to J. B. Harrington, June 18, 1964, MS 467, box 3, folder 46, Borlaug Papers, Iowa State University Library.

50. Borlaug to Harrington, June 18, 1964.

51. Borlaug, "Report of South American Trip, Nov 11–Dec 18, 1959," 56.

52. I. Narvaez and Norman Borlaug, "Progress Report: Accelerated Wheat

Improvement in West Pakistan, and the Revolution in Agriculture," March 30, 1966, p. 17, folder 191, box 28, series 323, RG 1.2, Projects, FA387, Rockefeller Foundation Records.

53. Borlaug, "A Brief Report on Progress Being Made by the Indian Coordinated Wheat Improvement Program," April 12, 1966, p. 11, coll. 01014, box 3, folder 5, Borlaug Papers, University of Minnesota Libraries.

54. Borlaug, "Brief Report on Progress," 11.

55. Hesser, *Man Who Fed the World*, 52.

56. S. H. Wittwer, "Observations, Impressions and Review of the Research Programs at the Environmental Research Laboratory University of Arizona, Tucson and the International Maize and Wheat Improvement Center (CIMMYT) Ciudad, Obregón, Sonora, Mexico April 12–17, 1969," p. 8, folder 49, box 9, series 105, RG 1.3, Projects, FA388, Rockefeller Foundation Records.

57. Food and Agriculture Organization of the United Nations (FAO), *Study on the Response of Wheat.*

58. FAO, *Study on the Response of Wheat.*

59. FAO, *Study on the Response of Wheat.*

60. Jones, "Green Revolution in Latin America."

61. Borlaug, "Breeding Wheat for High Yield," 586.

62. Oral history of Norman Borlaug, 192.

63. Norman Borlaug, "Green Revolution: Peace and Humanity," Nobel Lecture in Oslo, Norway, December 11, 1970, http://www.nobelprize.org/nobel_prizes/peace/laureates/1970/borlaug-lecture.html.

64. A phrase attributed to Stakman about Borlaug, but commonly used in descriptions of his personality.

65. Shiva, *Violence of the Green Revolution*, 86.

66. Stakman, Bradfield, and Mangelsdorf, *Campaigns against Hunger*, 283.

67. Borlaug, "Development of High Yielding, Broadly-Adapted Spring Wheat Varieties, and its Significance for Increasing World Food Production," n. d., MS 467, box 19, folder 32, Borlaug Papers, Iowa State University Library.

68. Oral history of Charles F. Krull, 1966, folder 8, box 17, RG 13, Oral Histories, FA119, Rockefeller Foundation Records.

69. Oral history of Charles F. Krull.

70. Krull et al., *Results of the Fourth Inter-American Spring Wheat Yield Nursery.*

71. Oral history of Norman Borlaug, 191.

72. Charles F. Krull, "The International Yield Nurseries and Their Implications," presentation to the Crop Quality Council, 1967, p. 3, MS 467, box 8, folder 26, Borlaug Papers, Iowa State University Library.

73. Krull et al., *Results of the Third Near East–American Spring Wheat Yield Nursery*, 10. Emphasis in the original. "Variety by location interactions" means that a variety performs differently depending on the location. The greater the

difference, the greater the "interaction" would be. This is a type of genotype by environment interaction.

74. Krull et al., *Results of the Third Near East–American Spring Wheat Yield Nursery*, 10.

75. Other FAO scientists in the Middle East with whom Krull and other RF scientists regularly corresponded included Juan Tessi, José Vallega, and J. B. Harrington.

76. Barnes, *Transnational Flows of Expertise*.

77. Krull, diary notes, April 5–May, 1966, p. 5, box 250, RG 12, Officers' Diaries F–L, FA392, Rockefeller Foundation Records, 5.

78. Krull, diary notes, April 5–May, 1966, 5.

79. Krull to Abdul Hafiz, June 23, 1966, coll. 01014, box 3, folder 5, Borlaug Papers, University of Minnesota Libraries.

80. Krull, diary notes, April–May, 1966, 12.

81. Krull, diary notes, April–May, 1966, 12.

82. C. L. Pan to Herbert K. Hayes, April 30, 1966, MS 467, box 4, folder 18, Borlaug Papers, Iowa State University Library.

83. Pan to Hayes, April 30, 1966.

84. Krull to Hafiz, June 23, 1966.

85. Hafiz to Krull, June 30, 1966, coll. 01014, box 3, folder 5, Borlaug Papers, University of Minnesota Libraries.

86. Krull, diary notes, April–May, 1966, 12.

87. Krull et al., *Results of the Fourth Inter-American Spring Wheat Yield Nursery*, 10.

88. Krull, "International Yield Nurseries and Their Implications," 3.

89. Krull, "International Yield Nurseries and Their Implications," 5.

90. Krull, "International Yield Nurseries and Their Implications," 5.

91. Krull, "International Yield Nurseries and Their Implications," 8.

92. Krull to R. R. Kalton, October 4, 1965, MS 467, box 3, folder 72, Borlaug Papers, Iowa State University Library.

93. Krull to Hafiz, June 23, 1966.

94. Krull to Byrd C. Curtis, January 5, 1967, MS 467, box 7, folder 8, Borlaug Papers, Iowa State University Library.

95. Krull to Curtis, January 5, 1967. Emphasis in the original.

96. Hafiz to Krull, June 30, 1966.

97. Krull to Hafiz, July 12, 1966, MS 467, box 5, folder 22, Borlaug Papers, Iowa State University Library.

98. Pan to Krull, March 20, 1966, MS 467, box 5, folder 35, Borlaug Papers, Iowa State University Library.

99. Pan to Krull, April 18, 1967, MS 467, box 7, folder 35, Borlaug Papers, Iowa State University Library.

100. Krull, "CFK's diary," May 28–June 8, 1967, pp. 3–4, box 250, RG 12,

Officers' Diaries F–L, FA392, Rockefeller Foundation Records. Emphasis in the original.

101. Robert W. Romig, "RWR Diary," February 20, 1968, folder 44, box 8, series 105, RG 1.3, Projects, FA388, Rockefeller Foundation Records.

102. Yates and Cochran, "Analysis of Groups of Experiments"; Horner and Frey, "Methods for Determining Natural Areas."

103. Finlay and Wilkinson, "Analysis of Adaptation."

104. Romagosa and Fox, "Genotype × Environment Interaction and Adaptation."

105. Evans, *Crop Evolution, Adaptation and Yield*, 163.

106. Orville Vogel to Borlaug, August 23, 1963, coll. 01014, box 3, folder 4, Borlaug Papers, University of Minnesota Libraries.

107. Borlaug to Robert D. Osler, January 6, 1964, MS 467, box 4, folder 45, Borlaug Papers, Iowa State University Library.

108. Borlaug to Keith W. Finlay, July 6, 1964, MS 467, box 3, folder 12, Borlaug Papers, Iowa State University Library.

109. Finlay to Borlaug, July 20, 1964, MS 467, box 19, folder 32, Borlaug Papers, Iowa State University Library.

110. Finlay to Borlaug, July 20, 1964.

111. Borlaug to Osler, July 6, 1964, MS 467, box 4, folder 35, Borlaug Papers, Iowa State University Library.

112. Louis P. Reitz to Osler, June 23, 1964, MS 467, box 4, folder 35, Borlaug Papers, Iowa State University Library.

113. Finlay to Borlaug, September 6, 1966, MS 467, box 19, folder 32, Borlaug Papers, Iowa State University Library.

114. Finlay to Borlaug, September 6, 1966.

115. Finlay to Borlaug, September 6, 1966.

116. Finlay to Borlaug, September 6, 1966.

117. Finlay to Borlaug, September 6, 1966.

118. Aronova, Baker, and Oreskes, "Big Science and Big Data in Biology."

119. Pistorius, *Scientists, Plants and Politics*; US National Committee for the IBP, *Preliminary Framework*, 29.

120. US National Committee for the IBP, *Preliminary Framework*, 29.

121. Otto H. Frankel, circular letter to members of working group on gene pools, August 3, 1966, MS 467, box 8, folder 12, Borlaug Papers, Iowa State University Library.

122. Borlaug to Wortman, August 24, 1966, MS 467, box 5, folder 51, Borlaug Papers, Iowa State University Library.

123. Frankel, "Adaptability of Crops."

124. Frankel to Borlaug, August 30, 1966, MS 467, box 8, folder 12, Borlaug Papers, Iowa State University Library.

125. CIMMYT, *Annual Report, 1972.*

126. Borlaug, "Wheat Breeding and Its Impact."

127. Finlay, "Significance of Adaptation in Wheat Breeding."

128. Borlaug and Gibler to Osler, August 1, 1968, MS 467, box 10, folder 9, Borlaug Papers, Iowa State University Library.

129. CIMMYT, *CIMMYT Report, 1968–69*, 7.

130. Reitz to Osler, June 23, 1964.

131. Stone and Glover, "Disembedding Grain."

CHAPTER 2: PROPER AGRONOMY

1. Abrol, "American Involvement in Indian Agricultural Research"; Anderson, "Cultivating Science as Cultural Policy"; Anderson, Levy, and Morrison, *Rice Science and Development Politics*; Lele and Goldsmith, "Development of National Agricultural Research Capacity"; Parayil, "Green Revolution in India"; Saha, "State Policy"; Subramanian, "Revisiting the Green Revolution."

2. Davis, *Late Victorian Holocausts*.

3. Parayil, "Green Revolution in India"; Perkins, *Geopolitics and the Green Revolution*.

4. Perkins, *Geopolitics and the Green Revolution*.

5. Knight, *Food Administration in India, 1939–47*.

6. Randhawa, *Agricultural Research in India*.

7. Anderson, "Cultivating Science as Cultural Policy."

8. Busch, *Universities for Development*.

9. Guha, *India after Gandhi*.

10. Parayil, "Green Revolution in India."

11. Howard and Howard, *Wheat in India*, 117.

12. Quoted in Saha, "State Policy," 95.

13. Government of India, *Intensive Agricultural Districts*, 82.

14. Indian Council of Agricultural Research [hereafter ICAR], Proceedings of the Meeting of the Advisory Board from the 3rd to the 5th January 1951, 1952, Indian Agricultural Research Institute Archives, New Delhi.

15. S. P. Kohli, "Wheat Varieties in India," 1968, p. 20, ICAR Technical Bulletin no. 18, folder 544, box 84, series 4, RG 6.7, Field Offices, New Delhi, FA396, Rockefeller Foundation Records, Rockefeller Archive Center, Tarrytown, NY.

16. ICAR, "Annual Report, 1951–52," 1954, Indian Agricultural Research Institute Archives.

17. Sikka and Jain, "Study of the Differential Response," 154.

18. IARI, Proceedings of the Meeting of the Advisory Board Held at New Delhi on 11 and 12 December, 1953, p. 32, Indian Agricultural Research Institute Archives.

19. Quoted in Guha, *India after Gandhi*, 206.

20. Guha, *India after Gandhi*.

21. Siegel, *Hungry Nation*.

22. Guha, *India after Gandhi.*

23. Guha, *India after Gandhi.*

24. Parayil, "Green Revolution in India."

25. Perkins, *Geopolitics and the Green Revolution*; Saha, "State Policy."

26. Cullather, *Hungry World.*

27. Saha, "State Policy."

28. Saha, "State Policy."

29. Saha, "State Policy."

30. Cullather, *Hungry World.*

31. Guha, *India after Gandhi*; Perkins, *Geopolitics and the Green Revolution.*

32. Latham, *Right Kind of Revolution.*

33. Williams, *Khanna Study.*

34. Quoted in Connelly, *Fatal Misconception*, 171.

35. Latham, *Right Kind of Revolution.*

36. Randhawa, *Agricultural Research in India.*

37. Abrol, "American Involvement in Indian Agricultural Research."

38. Abrol, "American Involvement in Indian Agricultural Research."

39. Rockefeller Foundation, *Director's Annual Report, January 15–September 15, 1957.*

40. Randhawa, *Agricultural Research.*

41. Quoted in Cullather, *Hungry World*, 199.

42. Government of India, *Intensive Agricultural Districts*, 10.

43. Sen, *Modernising Indian Agriculture.*

44. Dawson, Murphy, and Jones, "Decentralized Selection"; Gangopadhyaya and Sarker, "Influence of Rainfall Distribution."

45. Ulysses J. Grant and Edwin J. Wellhausen, "A Study of Corn Breeding and Production in India," 1955, folder 324, box 46, series 4, RG 6.7, Field Offices, New Delhi, FA396, Rockefeller Foundation Records.

46. M. W. Parker, E. E. Cheesman, R. L. Lovvorn, P. Maheshwari, K. Ramiah, O. B. Ross, and L. Sahai, first draft of "Report of the Agricultural Research Review Team," December 13, 1963, folder 552, box 58, series 464D, RG 1.2, Projects, FA387, Rockefeller Foundation Records.

47. ICAR, "Report Sub-Committee of the Botany Committee of ICAR," 1957, folder 258, box 39, series 4, RG 6.7, Field Offices, New Delhi, FA396, Rockefeller Foundation Records.

48. Anderson et al., *Science, Politics, and the Agricultural Revolution.*

49. R. Glenn Anderson, "Wheat Position Paper," 1970, folder 153, box 27, series 2, RG 6.7, Field Offices, New Delhi, FA396, Rockefeller Foundation Records.

50. Cullather, *Hungry World.*

51. Cullather, *Hungry World.*

52. Perkins, *Geopolitics and the Green Revolution.*

53. Norman E. Borlaug, "Indian Wheat Research Designed to Increase Wheat Production," April 11, 1964, p. 2, folder 546, box 84, series 4, RG 6.7, Field Offices, New Delhi, FA396, Rockefeller Foundation Records.

54. Borlaug, "Indian Wheat Research," 16.

55. Borlaug to M. S. Swaminathan, August 18, 1964, coll. 01014, box 3, folder 4, Norman E. Borlaug Papers, 1930–2006, University of Minnesota Libraries, Minneapolis, MN. Borlaug had met Anderson in 1958 at an international wheat research symposium; see "Memorial to Glenn Anderson and Keith Finlay," video, El Batán, Mexico, CIMMYT Publications Repository, http://hdl.handle.net/10883/4781.

56. Saha, "State Policy."

57. Ahlberg, "'Machiavelli with a Heart.'"

58. Cullather, *Hungry World*; Weisskopf, "Dependence and Imperialism in India."

59. Cullather, *Hungry World*.

60. Weisskopf, "Dependence and Imperialism in India."

61. Perkins, *Geopolitics and the Green Revolution*; Cullather, *Hungry World*; Patel, "Long Green Revolution."

62. Ahlberg, "'Machiavelli with a Heart.'"

63. Ahlberg, "'Machiavelli with a Heart'"; Cullather, *Hungry World*.

64. Patel, "Long Green Revolution," 14. Ahlberg, "'Machiavelli with a Heart,'" and Cullather, *Hungry World*, provide context on the political circumstances of this period in India. Perkins, *Geopolitics and the Green Revolution*, provides detail on changes in the Indian wheat research system.

65. Quoted in Frankel, "India's New Strategy of Agricultural Development," 694.

66. Abel, "Agriculture in India in the 1970s."

67. Cullather, *Hungry World*; Weisskopf, "Dependence and Imperialism in India."

68. Subramanian, "Revisiting the Green Revolution."

69. Orville L. Freeman, telegram from the embassy in Italy to the Department of State, November 26, 1965, US Department of State, Foreign Relations of the United States, 1964–1968: South Asia, Vol. XXV, Doc. 253, https://history.state.gov/historicaldocuments/frus1964-68v25/d253.

70. Rockefeller Foundation, *Towards the Conquest of Hunger*.

71. Perkins, *Geopolitics and the Green Revolution*.

72. Borlaug, "Indian Wheat Research," 18.

73. Borlaug, "A Brief Report on Progress Being Made by the Indian Coordinated Wheat Improvement Program," April 12, 1966, coll. 01014, box 3, folder 5, Borlaug Papers, University of Minnesota Libraries.

74. Cullather, *Hungry World*.

75. Rockefeller Foundation, *Annual Report, 1963–4*, 237.

76. ICAR, "A Proposal for the Initiation of a Coordinated Wheat Breeding Scheme on an All India Basis," 1965, p. 18, folder 547, box 84, series 4, RG 6.7, Field Offices, New Delhi, FA396, Rockefeller Foundation Records.

77. Minutes of the Fourth All India Wheat Research Workers' Seminar, p. 7, Directorate of Wheat Research Library, Karnal.

78. Raina, "Institutional Strangleholds."

79. Anderson, "Wheat Position Paper"; Rockefeller Foundation Indian Agricultural Program, "The Indian Agricultural Research Institute (Position Paper for IAP Review)," n.d., p. 14, folder 153, box 27, series 2, RG 6.7, Field Offices, New Delhi, FA396, Rockefeller Foundation Records.

80. Lele and Goldsmith, "Development of National Agricultural Research Capacity."

81. Pal, quoted in Krishna, *New Agricultural Strategy*, 6.

82. Proceedings of the 5th All India Wheat Research Workers' Conference, August 17–22, 1966, Jaipur, p. 226, Indian Agricultural Research Institute Archives, New Delhi; Indian Agricultural Research Institute [hereafter IARI], "IARI Annual Scientific Report, 1967," p. 16, Indian Agricultural Research Institute Archives; Proceedings of the 8th All India Wheat Research Workers' Conference, vol. 1, 1969, p. 174, Indian Agricultural Research Institute Archives.

83. Kohli, *Wheat Varieties in India*, 9.

84. Das and Jain, "Studies on Adaptation in Wheat," 83.

85. Swaminathan, "Impact of Dwarfing Genes," 61.

86. Proceedings of the 5th All India Wheat Research Workers' Conference, 243.

87. IARI, "IARI Annual Scientific Report, 1967," 16.

88. Roy and Murty, "Response to Selection," 481.

89. Mohammadi, "Breeding for Increased Drought Tolerance." Monneveux, Jing, and Misra, "Phenotyping for Drought Adaptation," 5.

90. Roy and Murty, "Selection Procedure," 516.

91. Proceedings of the 6th All India Wheat Research Workers' Workshop, vol. 2, 1967, p. 124, Indian Agricultural Research Institute Archives.

92. ICAR, "Proposal for the Initiation," 18; Proceedings of the 5th All India Wheat Research Workers' Conference, 25.

93. Government of India, Minutes of Varietal Release Committee Meeting, Draft, August 6, 1965, folder 527, box 82, series 4, RG 6.7, Field Offices, New Delhi, FA396, Rockefeller Foundation Records.

94. ICAR, "ICAR Annual Report 1965–66," 1966, p. 9, Indian Agricultural Research Institute Archives.

95. R. Glenn Anderson to Norman E. Borlaug, January 24, 1966, coll. 01014, box 3, folder 5, Borlaug Papers, University of Minnesota Libraries.

96. Sen, *Green Revolution in India*, 27.

97. Ralph W. Cummings, "RWC diary," February 5, 1966, folder 254, box 37, series 3, RG 6.7, Field Offices, New Delhi, FA396, Rockefeller Foundation Records.

98. Minutes of the Fourth All India Wheat Research Workers' Seminar, 14.

99. Minutes of the Fourth All India Wheat Research Workers' Seminar, 42.

100. Borlaug, "Brief Report on Progress"; Proceedings of the Seventh All India Wheat Research Workers' Workshop, 1968, folder 555, box 85, series 4, RG 6.7, Field Offices, New Delhi, FA396, Rockefeller Foundation Records.

101. Food and Agriculture Organization of the United Nations, *Fertilizer Use by Crop in India.*

102. Hopper to Borlaug, February 10, 1966, MS 467, box 5, folder 41, Norman E. Borlaug Papers. Special Collections Department, Iowa State University Library, Ames, IA.

103. Minutes of the Fourth All India Wheat Research Workers' Seminar, 14.

104. Oral history of Norman Borlaug, 1967, p. 196, coll. 01014, box 1, folder 5, Borlaug Papers, University of Minnesota Libraries; Government of India, Minutes of Varietal Release Committee Meeting.

105. Minhas and Srinivasan, "New Agricultural Strategy Analysed."

106. Minhas and Srinivasan, "New Agricultural Strategy Analysed," 24.

107. Subramanian, "Revisiting the Green Revolution," 53.

108. Wright to Panse, December 28, 1965, folder 780, box 112, series 6, RG 6.7, Field Offices, New Delhi, FA396, Rockefeller Foundation Records.

109. Wright to Panse, December 28, 1965.

110. Anderson, *Potentials for Improving Production Efficiency of Cereals*, 2.

111. Anderson, "RGA's Trip Report for India, November 1973," p. 84, folder 113, box 9, series 19, RG 6, Field Offices, Ankara (Turkey), FA399, Rockefeller Foundation Records.

112. Harwood, "Was the Green Revolution Intended," 318.

113. Harwood, "Was the Green Revolution Intended"; Cleaver, "Contradictions of the Green Revolution."

114. Report and Summary of the Agricultural Science Field Directors' Meeting, July 25–28, 1966, p. 8, folder 16, box 4, series 915, RG 3, Administration, Program, and Policy, FA112, Rockefeller Foundation Records.

115. Borlaug, "Brief Report," 11.

116. Borlaug to Albert H. Moseman, December 21, 1966, folder 179, box 26, series 323, RG 1.2, Projects, FA387, Rockefeller Foundation Records.

117. Proceedings of the 8th All India Wheat Research Workers' Conference, vol. 1, 1969, pp. 34–35, Indian Agricultural Research Institute Archives.

118. Proceedings of the 8th All India Wheat Research Workers' Conference, 36.

119. Government of India, Minutes of Varietal Release Committee Meeting.

120. Cummings, "RWC diary," March 17, 1964.

121. Borlaug to I. Narvaez and R. Glenn Anderson, October 15, 1967, MS 467, box 8, folder 2, Borlaug Papers, Iowa State University Library. Emphasis in the original.

122. Proceedings of the 6th All India Wheat Research Workers' Workshop, 17.

123. Proceedings of the 6th All India Wheat Research Workers' Workshop, 12.

124. Proceedings of the 8th All India Wheat Research Workers' Conference, 34.

125. Wright to Cummings, August 29, 1967, folder 529, box 82, series 4, RG 6.7, Field Offices, New Delhi, FA396, Rockefeller Foundation Records.

126. Proceedings of the 6th All India Wheat Research Workers' Workshop, 17.

127. Proceedings of the 6th All India Wheat Research Workers' Workshop, 17.

128. Kanwar, "Research for Effective Use," 214.

129. Oral history of Norman Borlaug, 196.

130. Proceedings of the 8th All India Wheat Research Workers' Conference, 174.

131. Subramanian, "Revisiting the Green Revolution."

132. Subramanian, "Revisiting the Green Revolution," 54.

133. Kohli, "Towards Further Rationalization in Plant Breeding."

134. Minhas and Srinivasan, "New Agricultural Strategy Analysed," 21.

135. Parikh, "HYV Fertilisers," A-5.

136. Keith W. Finlay to Borlaug, September 6, 1966, MS 467, box 19, folder 32, Borlaug Papers, Iowa State University Library.

137. Borlaug to Finlay, July 7, 1965, MS 467, box 10, folder 9, Borlaug Papers, Iowa State University Library.

138. Subramanian, "Revisiting the Green Revolution," 56.

139. Farmer, "'Green Revolution' in South Asian Ricefields."

140. Farmer, "'Green Revolution' in South Asian Ricefields," 308.

141. Anderson et al., *Science, Politics, and the Agricultural Revolution*; Cleaver, "Contradictions of the Green Revolution"; Frankel, *India's Green Revolution*; Griffin, *Political Economy of Agrarian Change*; Ladejinsky, "Green Revolution in Punjab."

142. Farmer, "'Green Revolution' in South Asian Ricefields"; Lewontin, "Green Revolution and the Politics"; Oasa, "International Rice Research Institute"; Saha, "State Policy"; Sen, *Green Revolution in India*.

143. R. W. Komer, Memorandum to President Johnson, March 27, 1966, US Department of

State, Foreign Relations of the United States, 1964–1968: South Asia, Vol. XXV, Doc. 306, http://history.state.gov/historicaldocuments/frus1964-68v25/d306.

144. Borlaug to Finlay, July 7, 1965.

CHAPTER 3: INDIAN WHEAT RESEARCH AFTER THE GREEN REVOLUTION

1. Dhanagare, "Green Revolution."

2. Cullather, *Hungry World*.

3. M. V. Rao, "Coordinator's Note," in "Results of the 1973–74 Coordinated Wheat Trials," Indian Council of Agricultural Research, 1974, p. viii, Directorate of Wheat Research Library, Karnal.

4. Cullather, *Hungry World*; Sen, *Green Revolution in India*.

5. Government of India, Directorate of Economics and Statistics, *Agricultural Statistics at a Glance 2015*.

6. Cullather, *Hungry World*.

7. Frankel, "India's New Strategy of Agricultural Development"; Ladejinsky, "Green Revolution in Punjab"; Ladejinsky, "Ironies of India's Green Revolution."

8. Edward Tenenbaum, "The Fertilizer Situation in India," March 21, 1975, p. 1-15, Folder ID 109921I, World Bank Group Archives, Washington, DC.

9. Abel, "Agriculture in India in the 1970s"; Chakravarti, "Green Revolution in India"; Cleaver, "Contradictions of the Green Revolution"; Sen, *Green Revolution in India*; Wade, "Green Revolution (I)."

10. Rockefeller Foundation Indian Agricultural Program, "The Indian Agricultural Program: Situation, Trends, Outlook," 1973, folder 4172, box 626, series 464, record group [RG] 1.6, Projects, Rockefeller Foundation Records Rockefeller Archive Center, Tarrytown, NY.

11. Nene, "Plant Protection"; Swaminathan, *Science and Sustainable Food Security*.

12. Singh and Pawar, *Trends in Wheat Breeding*.

13. Munshi, "Social Learning in a Heterogeneous Population."

14. M. V. Rao, "Coordinator's Note," in "Results of the 1974–75 Coordinated Wheat Trials," Indian Council of Agricultural Research, 1975, p. iv, Directorate of Wheat Research Library.

15. M. V. Rao, "Coordinator's Note," in "Results of the 1976–77 Coordinated Wheat Trials," Indian Council of Agricultural Research, 1977, p. 1.2, Directorate of Wheat Research Library.

16. J. P. Srivastava to R. Glenn Anderson, August 9, 1972, MS 467, box 17, folder 5, Norman E. Borlaug Papers, Special Collections Department, Iowa State University Library, Ames, IA.

17. Eugene E. Saari to R. Glenn Anderson, September 28, 1972, MS 467, box 16, folder 64, Borlaug Papers, Iowa State University Library.

18. Anderson, "RGA's Trip Report for India, November 1973," folder 113, box 9, series 19, RG 6, Field Offices, Ankara (Turkey), FA399, Rockefeller Foundation Records.

19. Tenenbaum, "Fertilizer Situation in India," p. 4-6; Sen, *Green Revolution in India*.

20. Govind, *Regional Perspectives in Agricultural Development.*

21. Tenenbaum, "Fertilizer Situation in India," p. 4-1.

22. Tenenbaum, "Fertilizer Situation in India," p. 2-13.

23. Tenenbaum, "Fertilizer Situation in India," p. 4-3.

24. A. B. Joshi, "Brief Resume on Crop Sciences Research," 1970, p. 1, folder 128, box 24, series 2, RG 6.7, Field Offices, New Delhi, FA396, Rockefeller Foundation Records.

25. Rao, "Coordinator's Note," 1975, iv.

26. Tenenbaum, "Fertilizer Situation in India," p. 3-5.

27. Rao, "Coordinator's Note," 1974, x.

28. A. B. Joshi, "Fertilizers and Agriculture in India," Dhiru Morarji Memorial Lecture, New Delhi, November 1, 1974, as cited in Tenenbaum, "Fertilizer Situation in India," p. 3-13.

29. Parikh, "HYV Fertilisers." A-5.

30. Parikh, "HYV Fertilisers." A-5.

31. Sagar, "Fertiliser Use Efficiency in Indian Agriculture."

32. Vaidyanathan, "HYV and Fertilisers," 1033.

33. Vaidyanathan, "HYV and Fertilisers," 1033.

34. Tenenbaum, "Fertilizer Situation in India," p. 3-7.

35. Desai, "Fertiliser Use in India."

36. J. P. Tandon, "Project Director's Report," in "Results of the 1987–88 Coordinated Wheat and Triticale Trials," Indian Council of Agricultural Research, 1988, Directorate of Wheat Research Library.

37. J. P. Tandon, "Project Director's Report," in "Results of the 1982–83 Coordinated Wheat and Triticale Trials," Indian Council of Agricultural Research, 1983, p. xix, Directorate of Wheat Research Library.

38. Wheat Project Directorate, "Notes Concerning Causes for Relatively Lower Productivity Levels Being Recorded in Research Trials Carried Out under AICWIP," report to Norman E. Borlaug, n.d., p. 2, MS 467, box 21, folder 14, Borlaug Papers, Iowa State University Library.

39. Borlaug to Rao and Tandon, January 6, 1987, MS 467, box 21, folder 14, Borlaug Papers, Iowa State University Library.

40. Haldore Hanson, Keith W. Finlay, R. Glenn Anderson, Norman E. Borlaug, and Ernest W. Sprague, "CIMMYT's Research and Training Activities outside Mexico," October 24, 1975, p. 11, CGSpace, https://hdl.handle.net/10947/375.

41. Anderson to Borlaug, December 20, 1969, MS 467, box 14, folder 38, Borlaug Papers, Iowa State University Library.

42. S. Sen to Guy B. Baird, February 17, 1967, folder 125, box 24, series 2, RG 6.7, Field Offices, New Delhi, FA396, Rockefeller Foundation Records.

43. Anderson to Borlaug, November 22, 1969, MS 467, box 11, folder 34, Borlaug Papers, Iowa State University Library.

44. Anderson, "RGA's Trip Report for India," 1.

45. Munshi, "Social Learning in a Heterogeneous Population."

46. Rao, "Coordinator's Note," 1974, x.

47. M. V. Rao, "Project Director's Report," in "Results of the 1979–80 Coordinated Wheat and Triticale Trials," Indian Council of Agricultural Research, 1980, p. 18, Directorate of Wheat Research Library.

48. CIMMYT, *Report on Wheat Improvement*, 1984.

49. Krishna et al., *Empirical Examination*.

50. Rockefeller Foundation Indian Agricultural Program, "The Indian Agricultural Research Institute (Position Paper for IAP Review)," n.d., folder 153, box 27, series 2, RG 6.7, Field Offices, New Delhi, FA396, Rockefeller Foundation Records.

51. Rockefeller Foundation Indian Agricultural Program, "Indian Agricultural Research Institute (Position Paper for IAP Review)."

52. Rockefeller Foundation Indian Agricultural Program, "All India Wheat Improvement Program," March 17–28, 1970, folder 153, box 27, series 2, RG 6.7, Field Offices, New Delhi, FA396, Rockefeller Foundation Records.

53. Rockefeller Foundation Indian Agricultural Program, "The Rockefeller Foundation in India, 1920–1972," June 1, 1972, folder 4171, box 626, series 464, RG 1.6, Projects, Rockefeller Foundation Records.

54. Rockefeller Foundation Indian Agricultural Program, "1973 Progress Report," December 1, 1973, folder 4174, box 627, series 464, RG 1.6, Projects, Rockefeller Foundation Records.

55. Naik, "Inaugural Address," 3.

56. Naik, "Inaugural Address," 3.

57. Naik, "Inaugural Address," 3.

58. Naik, "Inaugural Address," 1–2.

59. Sisodia, "Photoinsensitivity in Wheat Breeding," 173.

60. Sisodia, "Photoinsensitivity in Wheat Breeding," 173.

61. Sen, *Green Revolution in India*, 25.

62. Sen, *Green Revolution in India*, 25–26.

63. R. Glenn Anderson, memo, March 2, 1966, p. 2, folder 189, box 28, series 323, RG 1.2, Projects, FA387, Rockefeller Foundation Records.

64. Guy B. Baird to Sterling Wortman, February 4, 1970, folder 153, box 27, series 2, RG 6.7, Field Offices, New Delhi, FA396, Rockefeller Foundation Records.

65. Kanwar, "From Protective to Productive Irrigation."

66. Tandon, "Project Director's Report," 1988.

67. Wortman to Baird, September 15, 1969, folder 113, box 22, series 2, RG 6.7, Field Offices, New Delhi, FA396, Rockefeller Foundation Records. Emphasis in the original.

68. Ralph W. Cummings to Bill C. Wright, June 2, 1969, folder 207, box

33, series 2, RG 6.7, Field Offices, New Delhi, FA396, Rockefeller Foundation Records.

69. Baird to Delbert T. Myren, December 10, 1969, folder 789, box 113, series 6, RG 6.7, Field Offices, New Delhi, FA396, Rockefeller Foundation Records.

70. W. David Hopper to Edwin J. Wellhausen, December 2, 1969, folder 789, box 113, series 2, RG 6.7, Field Offices, New Delhi, FA396, Rockefeller Foundation Records.

71. Government of India, "All India Seminar on Dry Land Farming," January 1970, folder 127, box 24, series 2, RG 6.7, Field Offices, New Delhi, FA396, Rockefeller Foundation Records; Joshi, "Brief Resume on Crop Sciences Research," 6.

72. O. Starnes, diary notes, September 25–28, 1972, folder 4172, box 626, series 464, RG 1.6, Projects, Rockefeller Foundation Records.

73. M. V. Rao, "Project Director's Report," in "Results of the 1977–78 Coordinated Wheat and Triticale Trials," Indian Council of Agricultural Research, 1978, pp. 1.12–1.13, Directorate of Wheat Research Library. Emphasis in the original.

74. CIMMYT, *Enduring Designs for Change*.

75. M. V. Rao, "Project Director's Report," in "Results of the 1980–81 Coordinated Wheat and Triticale Trials," Indian Council of Agricultural Research, 1981, p. 1.39, Directorate of Wheat Research Library.

76. Tandon, "Project Director's Report," 1983, xix.

77. Swaminathan, *Wheat Revolution*, 35.

78. Tandon, "Project Director's Report," in "Results of the All India Coordinated Wheat and Triticale Varietal Trials 1992–93," Indian Council of Agricultural Research, 1993, Directorate of Wheat Research Library.

79. Quoted in Saha, "State Policy," 172.

80. Saha, "State Policy"; Shankar, "Towards a Paradigm Shift"; Witcombe, Virk, and Farrington, *Seeds of Choice*.

81. Easter, Bisaliah, Dunbar, "After Twenty-Five Years," 1203.

82. Yasin et al., "Genetic Improvement," 753.

83. Roy and Murty, "Selection Procedure," 516.

84. Tandon, "Project Director's Report," 1993, 1.31.

85. Sikka and Jain, "Study of the Differential Response," 154.

86. Sikka and Jain, "Study of the Differential Response," 154.

87. Kohli, "Towards Further Rationalization in Plant Breeding," 28.

88. R. Glenn Anderson to Charles F. Krull, October 24, 1966, MS 467, box 4, folder 73, Borlaug Papers, Iowa State University Library.

89. Keith W. Finlay, personal history record and application for travel grant, April 15, 1968, folder 683, box 69, series 464D, RG 1.2, Projects, FA387, Rockefeller Foundation Records.

90. Athwal and Singh, "Variability in Kangni—1"; Chandra, "Variability in

Gram"; Rao and Harinarayana, "Phenotypic Stability"; Ram, Jain, and Murty, "Stability of Performance."

91. Romagosa and Fox, "Genotype × Environment Interaction and Adaptation," reflect that the analysis required transferal of punch cards between institutions, which limited which research centers could perform the analysis.

92. Bhullar, Gill, and Khehra, "Stability Analysis"; Das and Jain, "Studies on Adaptation in Wheat"; Gupta, Virk, and Satija, "Quantitative Genetic Analysis in Wheat"; Luthra and Singh, "Comparison of Different Stability Models"; Mohan, Das, and Jain, "Note on Adaptation in Wheat"; Verma, Chahal, and Murty, "Limitations of Conventional Regression Analysis."

93. Bhullar, Gill, and Khehra, "Stability Analysis," 43–44.

94. Bhullar, Gill, and Khehra, "Stability Analysis," 44.

95. Mohan, Das, and Jain, "Note on Adaptation in Wheat,"1124.

96. Finlay to Borlaug, September 6, 1966, MS 467, box 19, folder 32, Borlaug Papers, Iowa State University Library.

97. Rao, "Project Director's Report," 1978, 1.10.

98. Newport et al., "Factors Constraining."

99. Nagarajan, Singh, and Tyagi, *Wheat Research Needs beyond 2000 AD.*

100. Tandon, "Project Director's Report," 1993, 1.46.

101. Sen, *Modernising Indian Agriculture*, 123.

102. Sen, *Modernising Indian Agriculture*, 46.

103. Government of India, *Report of the National Commission*, 151.

104. K. Kanungo and P. E. Naylor, "Elements of a National Research Project," ca. 1975, p. 5, folder 4175, box 627, series 464, RG 1.6, Projects, Rockefeller Foundation Records.

105. Kanungo and Naylor, "Elements of a National Research Project," 6.

106. Balaguru, Venkateswarlu, and Rajagopalan, "Implementation of World Bank."

107. Balaguru, Venkateswarlu, and Rajagopalan, "Implementation of World Bank."

108. World Bank, *Implementation Completion Report*, ii.

109. World Bank, *Report and Recommendations*, 13.

110. Kashyap and Mathur, "Ongoing Changes in Policy."

111. ENS Economic Bureau, "PM Asks States to Focus on Infra & Exports, Makes Pitch for Farmers," *Indian Express*, February 21, 2021, https://indianexpress.com/article/business/economy/narendra-modi-niti-aayog-farmers-infrastructure-exports-7197563/.

112. Tandon, "Project Director's Report," 1993, 1.39.

113. Singh, *Final Report of the Working Group*, 9.

114. Singh and Pawar, *Trends in Wheat Breeding*, 9.

115. Singh and Pawar, *Trends in Wheat Breeding*, 9.

116. Douthwaite, Keatinge, and Park, "Why Promising Technologies Fail."

117. Tandon and Rao, "Organisation of Wheat Research in India," 11.

118. Tandon and Rao, "Organisation of Wheat Research in India," 12.

119. J. P. Tandon, "Project Director's Report," in "Results of the 1983–84 Co-ordinated Wheat and Triticale Trials," Indian Council of Agricultural Research, 1984, p. 10, Directorate of Wheat Research Library.

120. Agrawal, "Development of Improved Varieties," 60–61.

121. Borlaug to Tandon, October 17, 1984, MS 467, box 21, folder 14, Borlaug Papers, Iowa State University Library.

122. Oasa, "International Rice Research Institute."

123. International Rice Research Institute, *IR8 and Beyond*, 1, as cited in Cullather, "Miracles of Modernization."

124. Oasa, "International Rice Research Institute," 250.

125. Munshi, "Social Learning in a Heterogeneous Population."

126. Subramanian, "Revisiting the Green Revolution," 71.

127. Oasa, "International Rice Research Institute," 251.

128. Harwood, *Europe's Green Revolution and Its Successors.*

129. Kalirajan and Shand, "Location Specific Research," 538.

130. Easter, Bisaliah, and Dunbar, "After Twenty-Five Years," 1203.

131. Easter, Bisaliah, and Dunbar, "After Twenty-Five Years," 1203.

132. Raina, "Institutional Strangleholds," 109.

133. Raina, "Institutional Strangleholds," 112.

CHAPTER 4: THE PERSISTENCE OF WIDE ADAPTATION IN INDIA

1. Bozeman and Sarewitz, "Public Values," 121.

2. Mishra, Ravindra, and Hesse, *Rainfed Agriculture.*

3. Desai et al., "Agricultural Policy Strategy, Instruments and Implementation," 43.

4. Raina, "Institutional Strangleholds."

5. Government of India, *Twelfth Five Year Plan*, 29.

6. This is also evident in the lack of baseline data on equity and sustainability metrics presented in the Five Year Plans, where most data are focused on crop-specific production and prices.

7. Krishna et al., *Farmer Access*; Majumdar et al., "Nutrient Management in Wheat."

8. Pathak, Panda, and Nayak, *Bringing Green Revolution to Eastern India.*

9. Basu, "Government Success, Failure of the Market," 491.

10. Freebairn, "Did the Green Revolution Concentrate Incomes," 1995.

11. Katyal and Mrutyunjaya, *CGIAR Effectiveness*, 34. The Operations Evaluation Department is an independent body that reports directly to the World Bank's board.

12. Harris and Orr, "Is Rainfed Agriculture," 84.

13. Harris and Orr, "Is Rainfed Agriculture," 93.

14. Raina, "Institutional Strangleholds,"113.

15. Raina, "Science for a New Agricultural Policy," 251.

16. Rao, *Food, Nutrition and Poverty in India*, 137.

17. Ramadas, Kumar, and Singh, "Wheat Production in India."

18. Raina, "Science for a New Agricultural Policy," 252.

19. Patel, "Long Green Revolution," 47.

20. Eliazar Nelson, Ravichandran, and Antony, "Impact of the Green Revolution."

21. Desai et al., *Undernutrition and Public Policy in India*.

22. Government of India, *National Workshop*, 1.

23. Spitz, "Green Revolution Re-examined in India."

24. Glover and Poole, "Principles of Innovation."

25. Davis, *Late Victorian Holocausts*.

26. Chatterjee and Kapur, "Understanding Price Variation."

27. Narayanan, "Food Security in India"; Sandip Das, "PDS Reforms," press release, June 27, 2016, Government of India Press Information Bureau, https://pib.gov.in/newsite/PrintRelease.aspx?relid=146019.

28. Ceccarelli, "Wide Adaptation"; Mishra, Ravindra, and Hesse, *Rainfed Agriculture*; Raina, "Institutional Strangleholds"; Snapp, "Mini-review on Overcoming"; Witcombe, Dirk, and Farrington, *Seeds of Choice*.

29. Krishna, Spielman, and Veettil, "Exploring the Supply and Demand."

30. Nagarajan, "Can India Produce Enough Wheat."

31. Babu et al., "State of Agricultural Extension Reforms," 162.

32. Over 70 percent of India's spring bread wheat varieties released from 1988 to 2002 were a direct selection from a CIMMYT variety or had a CIMMYT variety as a parent line. Lantican, Dubin, and Morris, *Impacts of International Wheat Breeding Research*.

33. Krishna, Spielman, and Veettil, "Exploring the Supply and Demand."

34. Traxler and Byerlee, "Linking Technical Change to Research Effort"; Krishna, Spielman, and Veettil, "Exploring the Supply and Demand"; Krishna et al., *Farmer Access*.

35. Krishna, Spielman, and Veettil, "Exploring the Supply and Demand."

36. Krishna et al., *Empirical Examination*, 18.

37. Bellon and Risopoulos, "Small-Scale Farmers."

38. Witcombe, Virk, and Farrington, *Seeds of Choice*.

39. Singh and Agarwal, "Improving Efficiency."

40. Witcombe, Virk, and Farrington, *Seeds of Choice*, give the example that shorter-duration varieties tend to have lower yields than the "check" varieties used as a goalpost, so they often fail to clear that hurdle despite the potential benefits or more reliable yields in a shorter growing season.

41. Nelson, Coe, and Haussmann, "Farmer Research Networks," 8.

42. Government of India, *National Workshop*, 1.

43. Takeshima, "Geography of Plant Breeding Systems," 67.

44. Government of India, *National Workshop*, 1.

45. Acharya and Das, "Revitalising Agriculture in Eastern India"; Mishra, Ravindra, and Hesse, *Rainfed Agriculture.*

46. Acharya and Das, "Revitalising Agriculture in Eastern India," 105.

47. Nagarajan, Singh, and Tgyagi, *Wheat Research Needs beyond 2000 AD*, 99.

48. N. N. Srinivas, "Crocodile Tears for Monsoon Rains Won't Do," *Economic Times of India*, July 16, 2012, https://economictimes.indiatimes.com/nidhi-nath-srinivas/crocodile-tears-for-monsoon-rains-wont-do/articleshow/14972567.cms.

49. Joshi et al., "Wheat Improvement in India."

50. Nagarajan, Singh, and Tgyagi, *Wheat Research Needs beyond 2000 AD*, 22.

51. MacDonald et al., "Groundwater Quality."

52. Yadav et al., "Wheat Production in India."

53. Shiferaw et al., "Crops that Feed the World 10."

54. Shiferaw et al., "Crops that Feed the World 10."

55. Yapa, "What Are Improved Seeds."

56. Bänziger and Cooper, "Breeding for Low Input Conditions"; Brancourt-Hulmel et al., "Indirect versus Direct Selection"; Calhoun et al., "Choosing Evaluation Environments"; Cooper et al., "Selection Strategy"; Cooper et al., "Wheat Breeding Nurseries."

57. For Indian wheat production, "germplasm improvement is still paramount," according to Rajbir Yadav et al., and until recently, breeding was focused on general conditions and not farming systems or specific climatic concerns such as heat stress. Yadav et al., "Wheat Production in India," 166.

58. *Innovation* has been defined by one group of researchers as "new and improved products and processes, new organizational forms, the application of existing technology to new fields, the discovery of new resources, and the opening of new markets." Niosi et al., "National Systems of Innovation," 209.

59. Glover and Poole, "Principles of Innovation."

60. Biggs, "Multiple Source of Innovation Model"; Hall et al., "New Agendas"; Klerkx and Leeuwis, "Matching Demand and Supply"; Röling, "Pathways for Impact"; Thompson and Scoones, "Addressing the Dynamics of Agri-Food Systems."

61. "According to the linear model, innovation happens in the following way: basic or fundamental research contributes to a general pool of knowledge; that pool of knowledge provides a resource for engineers or other innovators, who then apply it to create products that increase productivity, drive economic growth, enhance military power, and otherwise enrich lives and benefit society. This model assumes that advances in knowledge are by and large beneficial to

society, and that the benefits are both automatic and unpredictable. It also assumes a unidirectional flow of knowledge that privileges basic research above applied as the originator of all scientific benefit." Meyer, "Public Values Failures," 64.

62. Cash, Borck, and Patt, "Countering the Loading-Dock Approach," 484. Others have called this the "over the wall" model. In this model, researchers or developers assume that "the technology is complete and ready to use" and "the users are technically skilled enough to use it without help." Douthwaite, *Enabling Innovation*, 28.

63. A recent report by USAID found that the handoff between research and private sector actors was not a useful way to scale innovations (*scaling* is a term meaning the expanded adoption of an innovation, typically measured by the number of users). The authors found that "the 'handoff' model makes insufficient allowance for the need to modify and adapt technologies iteratively in response to market responses." US Agency for International Development, *Synthesis Report*, 35. In terms of scaling, "linear scaling models like this have increasingly been found to be inappropriate where fine scale variation amongst smallholders operating complex farming systems makes the performance of technologies very different for different farmers." Sinclair and Coe, "Options by Context Approach," 2. Technologies can be more easily scaled across larger farms, which tend to be more uniform due to irrigation and mechanization. The linear model of innovation promotes the "replication of simplicity rather than local adaptation and investigation of complexity." Nelson, Coe, and Haussmann, "Farmer Research Networks," 7.

64. Douthwaite, *Enabling Innovation*.

65. This is shown in statements such as "basic and strategic research that generate 'spillovers' broadly." Mruthyunjaya and Ranjitha, "Indian Agricultural Research System," 1095.

66. Biggs, "Multiple Source of Innovation Model."

67. Babu et al., "State of Agricultural Extension Reforms."

68. Raina et al., *Agricultural Innovation Systems*.

69. Sumberg and Reece, "Agricultural Research," 300.

70. K. S. Sudhi, "Agricultural Varsity Fails to Reap Success," *Hindu*, March 27, 2013, https://www.thehindu.com/news/cities/Kochi/agriculture-varsity-fails-to-reap-success/article4551944.ece.

71. This is one reason that the American land grant model of agricultural research and education did not translate well in Mexico and other countries. See Fitzgerald, "Exporting American Agriculture."

72. Raina, "Science for a New Agricultural Policy."

73. Bozeman and Sarewitz, "Public Values," 122. See also Meyer, "Public Values Failures" and Trouiller et al., "Drug Development for Neglected Diseases."

74. Babu et al., "State of Agricultural Extension Reforms."

75. Hall et al., "Public-Private Sector Interaction."

76. Raina, "But Why," 2.

77. Mruthyunjaya and Ranjitha, "Indian Agricultural Research System," 1095.

78. Raina et al., *Agricultural Innovation Systems.*

79. Nagarajan, "Can India Produce Enough Wheat," 1468.

80. Kumar, *Rethinking Revolutions.*

81. Subramanian, "Revisiting the Green Revolution," 56.

82. Harwood, *Europe's Green Revolution and Its Successors.*

83. Röling, "Pathways for Impact," 90.

84. Boru Douthwaite et al. have noted that "the ability of rural communities to innovate in the face of change is an important aspect of their resilience and sustainability." Douthwaite et al., "Why Promising Technologies Fail," 59. Dominic Glover and Nigel Poole wrote that "the countries of South Asia need more innovation" to meet their sustainable development goals for food and nutrition security. They continued, "However, they will also benefit from intelligent reflection about what innovation means, the directions it should take, and its risks and downsides alongside its benefits." Glover and Poole, "Principles of Innovation," 63–64, 64.

85. Glover and Poole, "Principles of Innovation."

86. Schut et al., "Innovation Platforms."

87. Schut et al., "Innovation Platforms," 539.

88. Sheldrake, "Setting Innovation Free in Agriculture," 26.

89. See Visser, "Down to Earth," 47, on the role of high modernity in twentieth-century agricultural research.

90. Yapa, "What Are Improved Seeds," 271.

91. Chambers, *Paradigm Shifts*, 2.

92. Scott, *Seeing Like a State.*

93. Harwood, *Europe's Green Revolution and Its Successors.*

94. Arora et al. "Control, Care, and Conviviality."

95. Kumar, *Water Policy Science and Politics.*

96. Raina et al., *Reviving Knowledge.*

97. Raina, "But Why," 5.

98. Glover and Poole, "Principles of Innovation," 71.

99. Glover and Poole use the example of Golden Rice, which is often touted by biotechnology advocates as a simple solution to vitamin A deficiency. They show, however, that to successfully implement Golden Rice technology would require reconfiguring agricultural value chains, which is much more complex than other nutritional fortification options.

100. Joshi et al., "Highly Client-Oriented Breeding."

101. Harwood, *Europe's Green Revolution and Its Successors*, 130.

102. Sumberg and Reece, "Agricultural Research," 296.

103. Sumberg and Reece, "Agricultural Research," 311.

104. Sumberg and Reece, "Agricultural Research," 311.

105. Klerkx and Leeuwis, "Matching Demand and Supply," 468.

106. Nelson, Coe, and Haussmann, "Farmer Research Networks."

107. Sumberg and Reece, "Agricultural Research," 297.

108. Descheemaeker et al., "Which Options Fit Best."

109. Descheemaeker et al., "Which Options Fit Best," 184.

110. Descheemaeker et al., "Which Options Fit Best," 185.

111. Sinclair and Coe, "Options by Context Approach."

112. Anderson et al., *Science, Politics, and the Agricultural Revolution.*

113. International Food Policy Research Institute, *Global Food Policy Report 2014–2015*, 29.

CHAPTER 5: CHALLENGES TO WIDE ADAPTATION IN INTERNATIONAL AGRICULTURAL RESEARCH

1. Harwood, "Could the Adverse Consequences."

2. Harwood, "Could the Adverse Consequences," 514.

3. CIMMYT, "A Proposal for Regional Coordination of the Mediterranean and Near East Cereal Programs," 1971, p. 56, folder 98, box 8, series 19, record group [RG] 6, Field Offices, Ankara (Turkey), FA399, Rockefeller Foundation Records, Rockefeller Archive Center, Tarrytown, NY.

4. Norman Borlaug to Edwin J. Wellhausen, November 10, 1967, folder 43, box 8, series 105, RG 1.3, Projects, FA388, Rockefeller Foundation Records.

5. CIMMYT, *Annual Report, 1972.*

6. Wellhausen, "Urgency of Accelerating Production," 6.

7. Wellhausen, "Urgency of Accelerating Production," 7.

8. Wellhausen, "Urgency of Accelerating Production," 7.

9. Lancaster, *Foreign Aid.*

10. Jefferson, "How Are Accountability Standards Implemented," 459.

11. Latham, *Right Kind of Revolution.*

12. Latham, *Right Kind of Revolution*, 169.

13. Ozgediz, *CGIAR at 40.*

14. Pistorius, *Scientists, Plants and Politics.*

15. CIMMYT, *Annual Report, 1972.*

16. Sterling Wortman to Guy B. Baird, September 15, 1969, folder 113, box 22, series 2, RG 6.7, Field Offices, New Delhi, FA396, Rockefeller Foundation Records; Ralph W. Cummings to Bill C. Wright, June 2, 1969, folder 207, box 33, series 2, RG 6.7, Field Offices, New Delhi, FA396, Rockefeller Foundation Records.

17. CGIAR Secretariat, "1978 CGIAR Integrative Report," September 1, 1978, p. 13, CGSpace, https://hdl.handle.net/10947/5410.

18. Harwood, *Europe's Green Revolution and Its Successors.*

19. CIMMYT, "What Is CIMMYT?" December, 1972, folder 105, box 8, se-

ries 19, RG 6, Field Offices, Ankara (Turkey), FA399, Rockefeller Foundation Records.

20. CIMMYT, "What Is CIMMYT."

21. Brooks et al., *Environmental Change.*

22. Haldore Hanson, Keith W. Finlay, R. Glenn Anderson, Norman E. Borlaug, and Ernest W. Sprague, "CIMMYT's Research and Training Activities outside Mexico," October 24, 1975, p. 11, CGSpace, https://hdl.handle .net/10947/375.

23. Quoted in Anderson, *Wheat, Triticale and Barley Seminar,* 23.

24. Oasa, "Political Economy of International Agricultural Research," 33.

25. Harwood, "Could the Adverse Consequences," 511.

26. Borlaug, "Green Revolution," 124.

27. Freebairn, "Did the Green Revolution Concentrate Incomes," 268.

28. Subramanian, "Revisiting the Green Revolution," 69–70.

29. Subramanian, "Revisiting the Green Revolution," 71.

30. Sivaraman, "Scientific Agriculture Is Neutral to Scale," 13.

31. Sivaraman, "Scientific Agriculture Is Neutral to Scale," 13.

32. Griffin, *Political Economy of Agrarian Change.*

33. Harwood, "Peasant Friendly Plant Breeding."

34. Harwood, "Peasant Friendly Plant Breeding."

35. Cotter, *Troubled Harvest,* 196.

36. Myren, "Rockefeller Foundation Program," 444.

37. Myren, "Rockefeller Foundation Program."

38. Myren, "Rockefeller Foundation Program."

39. Cotter, *Troubled Harvest*; Lewontin, "Green Revolution"; Matchett, "Untold Innovation."

40. Matchett, "Untold Innovation."

41. Arce, "Bureaucratic Conflict and Public Policy."

42. Edwin J. Wellhausen to Sterling Wortman, October 14, 1966, folder 73, box 13, series 105, RG 1.3, Projects, FA388, Rockefeller Foundation Records.

43. Rockefeller Foundation, Conference on Social Science Research in Rural Development, Agenda and Conference Papers, April 29–30, 1975, p. 4, University of Florida Digital Collections, http://ufdc.ufl.edu/UF00023152/00001.

44. Forrest F. Hill to Borlaug, June 29, 1967, MS 467, box 3, folder 25, Norman E. Borlaug Papers, Special Collections Department, Iowa State University Library, Ames, IA.

45. Redclift, "Production Programs for Small Farmers."

46. Redclift, "Production Programs for Small Farmers," 552.

47. Rockefeller Foundation, Conference on Social Science Research, 7.

48. Redclift, "Production Programs for Small Farmers."

49. CIMMYT, *Puebla Project.*

50. Wortman, notes on the Puebla Project and "Plan Maiz" of the State of

Mexico, October 26, 1971, folder 77, box 14, series 105, RG 1.3, Projects, FA388, Rockefeller Foundation Records.

51. Delbert T. Myren, "The Lessons of Puebla and the Potential of the Puebla Approach," 1972, p. 10, folder 78, box 14, series 105, RG 1.3, Projects, FA388, Rockefeller Foundation Records.

52. Myren, "Lessons of Puebla," 10.

53. Myren, "Lessons of Puebla," 12.

54. Myren, "Lessons of Puebla," 14.

55. Redclift, "Production Programs for Small Farmers."

56. Dahlberg, *Beyond the Green Revolution*, 54–55.

57. R. Bruner, "CIMMYT Third Quarterly Report, 1969: The Puebla Project," 1969, pp. 20, 25, folder 75, box 14, series 105, RG 1.3, Projects, FA388, Rockefeller Foundation Records.

58. R. Hertford, "Plan Puebla: Transferable and Generalizable?" August 17, 1970, folder 14, box 76, series 105, RG 1.3, Projects, FA388, Rockefeller Foundation Records; Myren, "Lessons of Puebla."

59. Kronstad, *Global Report*, 32.

60. Breth, "Turkey's Wheat Research and Training Project"; Dworkin, *Viking in the Wheat Field*.

61. Breth, "Turkey's Wheat Research and Training Project." OSU was chosen as a partner by Borlaug's longtime collaborator Orville Vogel, who personally selected wheat scientists at OSU after surveying Turkey for USAID. Duncan, "Wheat Dreams."

62. CIMMYT Board of Directors, "CIMMYT Agenda: IV Meeting of Board of Directors, Sept 25–26, 1969, Mexico," folder 21, box 4, series 105, RG 1.3, Projects, FA388, Rockefeller Foundation Records.

63. Arthur Klatt, Report on Trip to Turkey, May 25–June 17, 1971, folder 10, box 1, series 19, RG 6, Field Offices, Ankara (Turkey), FA399, Rockefeller Foundation Records, 10; Klatt to N. Eustatin, September 21, 1972, folder 31, box 2, series 19, RG 6, Field Offices, Ankara (Turkey), FA399, Rockefeller Foundation Records.

64. Breth, "Turkey's Wheat Research and Training Project," 9.

65. William C. Wright, "The Mid-East Wheat Project," October 5, 1971, folder 10, box 1, series 19, RG 6, Field Offices, Ankara (Turkey), FA399, Rockefeller Foundation Records.

66. Ralph W. Cummings, H. A. Rodenhiser, and John W. Gibler, "The Cooperative Wheat Research Program in Turkey," 1968, folder 1, box 1, series 19, RG 6, Field Offices, Ankara (Turkey), FA399, Rockefeller Foundation Records.

67. Klatt to Zoltan Barabas, n.d., folder 74, box 6, series 19, RG 6, Field Offices, Ankara (Turkey), FA399, Rockefeller Foundation Records.

68. Duncan, "Wheat Dreams."

69. Michaels, "Response of the 'Green Revolution.'"

70. Charles K. Mann to A. H. Bunting, September 8, 1977, folder 42, box 3, series 19, RG 6, Field Offices, Ankara (Turkey), FA399, Rockefeller Foundation Records.

71. Ceccarelli, "Adaptation to Low/High Input Cultivation"; Simmonds, "Selection for Local Adaptation"; Mann to Bunting, September 8, 1977.

72. Annicchiarico, *Genotype × Environment Interactions*.

73. Klatt, "Breeding for Yield Potential," 104.

74. Klatt, "Breeding for Yield Potential," 104–5.

75. Manassah and Briskey, *Advances in Food-Producing Systems*, 77.

76. Breth, "Turkey's Wheat Research and Training Project," 13.

77. Mann, "Packages of Practices."

78. CIMMYT, *CIMMYT Report 1980,* vi.

79. Villareal and Klatt, *Wheats*.

80. CIMMYT, *CIMMYT Report on Wheat Improvement*, vi.

81. Braun et al., "Turkish Wheat Pool."

82. Keser et al., "Genetic Gains in Wheat in Turkey," explains that even when the parent materials contained the Green Revolution dwarfing genes, those were not selected in Turkish plant breeding experiments under rainfed conditions. Other studies also found the Green Revolution semidwarfing traits are "not beneficial in the Mediterranean environment, where high temperatures and drought are commonly encountered." Yediay et al., "Allelic State at the Major Semi-dwarfing Genes," 424. Dwarfing genes derived from Russian and Italian varieties may be better adapted to the Turkish climate.

83. Mazid et al., "Measuring the Impact of Agricultural Research."

84. Albert H. Moseman, "AHM 1963 diary," February 25, 1963, box 342, RG 12, Officers' Diaries M–R, FA393, Rockefeller Foundation Records, 71.

85. CIMMYT, *CIMMYT Report, 1967–68*, 62.

86. J. V. Remenyi, "ALAD: An Evaluation," 1978, unpublished reports, #007493, Ford Foundation Records, Rockefeller Archive Center, Tarrytown, NY.

87. Remenyi, "ALAD: An Evaluation," 21.

88. CIMMYT, *CIMMYT Report, 1967–68*; CIMMYT, *CIMMYT Report, 1968–69*.

89. CIMMYT, *CIMMYT Report, 1968–69*.

90. Remenyi, "ALAD: An Evaluation."

91. Baum, *Partners against Hunger*.

92. Ford Foundation, "The Arid Lands Agricultural Development (ALAD) Program in the Middle East," December, 1973, folder 56, box 5, series 19, RG 6, Field Offices, Ankara (Turkey), FA399, Rockefeller Foundation Records; Remenyi, "ALAD: An Evaluation," 1.

93. Ford Foundation, "Arid Lands Agricultural Development (ALAD) Program," 2.

94. Remenyi, "ALAD: An Evaluation," 2.

95. Remenyi, "ALAD: An Evaluation," 61.

96. Remenyi, "ALAD: An Evaluation," 61.

97. Remenyi, "ALAD: An Evaluation," 2–3.

98. Remenyi, "ALAD: An Evaluation," 13.

99. Remenyi, "ALAD: An Evaluation."

100. R. Glenn Anderson, "RGA's Trip Report for India, November 1973," folder 113, box 9, series 19, RG 6, Field Offices, Ankara (Turkey), FA399, Rockefeller Foundation Records.

101. "Points Discussed between the Wheat Program Staff of the Arid Lands Agricultural Development Program and Dr. Abdul Hafiz, Regional Consultant, F.A.O.," May 31, 1971, folder 51, box 4, series 19, RG 6, Field Offices, Ankara (Turkey), FA399, Rockefeller Foundation Records.

102. Remenyi, "ALAD: An Evaluation," 61.

103. Remenyi, "ALAD: An Evaluation," 62.

104. CIMMYT, "Proposal for Regional Coordination."

105. Finlay to Wright, September 20, 1971, folder 100, box 8, series 19, RG 6, Field Offices, Ankara (Turkey), FA399, Rockefeller Foundation Records.

106. Skilbeck et al., *Research Review Mission*, 2.

107. Baum, *Partners against Hunger.*

108. Skilbeck et al., *Research Review Mission.*

109. Baum, *Partners against Hunger*, 86.

110. Skilbeck et al., *Research Review Mission*, 14.

111. Norman Borlaug, R. Glenn Anderson, Keith W. Finlay, and Haldore Hanson, CIMMYT comments on Proposed Agricultural Research Institute for the Mediterranean Region, CGIAR Online Repository, 1973, http://hdl.handle.net/10947/373.

112. Borlaug et al., CIMMYT comments.

113. Borlaug et al., CIMMYT comments, 6.

114. Borlaug et al., CIMMYT comments, 13. Emphasis in the original.

115. Borlaug et al., CIMMYT comments, 16.

116. Borlaug et al., CIMMYT comments, 38.

117. CGIAR, "Proposed Research Center in Middle East Region," January 3, 1975, folder 703, box 21, series IV A, FA624, David Bell files 1966–81, Ford Foundation Records.

118. Baum, *Partners against Hunger.*

119. International Development Research Centre (IDRC), News no. 24, 1976, folder 86, box 7, series 19, RG 6, Field Offices, Ankara (Turkey), FA399, Rockefeller Foundation Records.

120. Remenyi, "ALAD: An Evaluation."

121. M. B. Russell, "ICARDA: A Preliminary View," January 1975, p. 31, unpublished reports #008284, Ford Foundation Records.

122. Russell, "ICARDA," 1.

123. Xynias et al., "Durum Wheat Breeding in the Mediterranean Region," 432.

124. Russell, "ICARDA," 31.

125. Russell, "ICARDA," 31.

126. CGIAR Technical Advisory Committee [hereafter TAC], *TAC Statement*, 2.

127. CGIAR TAC, *TAC Statement*, 2–3.

128. Lantican et al., *Impacts of International Wheat Improvement Research*.

129. Xynias et al., "Durum Wheat Breeding in the Mediterranean Region."

130. Xynias et al., "Durum Wheat Breeding in the Mediterranean Region."

131. Ekboir, *CGIAR Crossroads*.

132. Dahlberg, "Ethical and Value Issues," 107.

133. Simmonds, "Genotype (*G*), Environment (*E*) and *GE* Components," 362.

134. Ceccarelli, "Wide Adaptation," 203.

135. Simmonds, "Selection for Local Adaptation"; Ceccarelli et al., "Genotype by Environment Interaction."

136. Simmonds, "Selection for Local Adaptation," 36.

137. Ceccarelli et al., "Genotype by Environment Interaction," 4.

138. Hill, Becker, and Tigerstedt, *Quantitative and Ecological Aspects*, 209.

139. Ceccarelli, "Wide Adaptation."

140. Ceccarelli, "Adaptation to Low/High Input Cultivation."

141. Simmonds, "Selection for Local Adaptation."

142. Tigerstedt, "Adaptation, Variation and Selection," 171.

143. Ceccarelli, "Adaptation to Low/High Input Cultivation," 212.

144. Such as the International Crops Research Institute for the Semi-Arid Tropics (ICRISAT) and the International Institute of Tropical Agriculture (IITA). Hill, Becker, and Tigerstedt, *Quantitative and Ecological Aspects*.

145. CIMMYT, *Enduring Designs for Change*, 13.

146. CIMMYT, *Enduring Designs for Change*, 13.

147. Braun, Rajaram, and van Ginkel, "CIMMYT's Approach"; Calhoun et al., "Choosing Evaluation Environments."

148. Calhoun et al., "Choosing Evaluation Environments," 677.

149. Calhoun et al., "Choosing Evaluation Environments."

150. Hurd, "Method of Breeding"; Hurd, "Phenotype and Drought Tolerance in Wheat."

151. Bidinger, Mahalakshmi, and Rao, "Assessment of Drought Resistance," 37.

152. Laing and Fischer, "Adaptation of Semidwarf Wheat Cultivars"; Fischer and Maurer, "Drought Resistance in Spring Wheat Cultivars." Fischer was affiliated with Australia's Commonwealth Scientific and Industrial Research Organisation and later director of CIMMYT's wheat program.

153. Rahman, Hayes, and Foster, *Manual of Wheat Breeding Procedures*, 80.

154. Rosielle and Hamblin, "Theoretical Aspects."

155. Ceccarelli, "Yield Potential."

156. One might recall Srivastava from chapter 3: indeed, this is the very same Srivastava from India who became a prominent researcher of dryland and rainfed agriculture.

157. Quoted in Ceccarelli, "Yield Potential," 226.

158. CIMMYT, *Report on Wheat Improvement* (1984), 7.

159. Hazell, *Summary Proceedings*.

160. Rajaram, Borlaug, and Van Ginkel, "CIMMYT International Wheat Breeding."

161. Gourdji et al., "Assessment of Wheat Yield Sensitivity."

162. Reynolds, *Climate Change and Crop Production*, 126.

163. CGIAR-IEA, *Evaluation of CGIAR Research Program*, ix.

164. CGIAR-IEA, *Evaluation of CGIAR Research Program*, 17.

165. CIMMYT and ICARDA, *CRP 3.1: Wheat.*

166. CGIAR-IEA, *Evaluation of CGIAR Research Program*, 23.

167. CIMMYT and ICARDA, *CRP 3.1: Wheat.*

168. Glaeser, *Green Revolution Revisited.*

169. Glaeser, *Green Revolution Revisited*; Harwood, *Europe's Green Revolution and Its Successors.*

170. Waines and Ehdaie, "Domestication and Crop Physiology."

171. Krishna et al., "Assessing Technological Change," 7.

CONCLUSION: THE LEGACY OF WIDE ADAPTATION IN INTERNATIONAL AGRICULTURAL DEVELOPMENT

1. For more information, see Colorado Wheat, "Why Is the Wheat Genome So Complicated?" November 15, 2013, https://coloradowheat.org/2013/11/why-is-the-wheat-genome-so-complicated/.

2. Shiferaw et al., "Crops that Feed the World," 10.

3. Vaidyanathan, "HYV and Fertilisers," 1033.

4. Subramanian, "Revisiting the Green Revolution," 43.

5. Sivaraman, "Scientific Agriculture Is Neutral to Scale," 16.

6. Piesse, *Hunger amid Abundance.*

7. Piesse, *Hunger amid Abundance.*

8. US Department of Agriculture Foreign Agricultural Service, *Grain and Feed Annual.*

9. Data obtained from Food and Agriculture Organization of the United Nations, *FAOSTAT Statistical Database*, https://www.fao.org/faostat/en/#data.

10. Kapil Subramanian has done so for irrigation.

11. Sumberg, Keeney, and Dempsey, "Public Agronomy."

12. Sumberg, Keeney, and Dempsey, "Public Agronomy," 1597.

13. Borlaug, "Green Revolution," 123.

14. Harwood, "Could the Adverse Consequences," 520.

15. Mann, *Wizard and the Prophet*, 122.

16. Oral history of Norman Borlaug, 1967, p. 42, coll. 01014, box 1, folder 5, Norman E. Borlaug Papers, 1930–2006, University of Minnesota Libraries, Minneapolis, MN.

17. Sumberg, Keeney, and Dempsey, "Public Agronomy," 1597.

18. Robert Chambers, "Personal Reflections on the Green Revolutions Narrative and Myths," Institute of Development Studies, May 24, 2019, https://www.ids.ac.uk/opinions/personal-reflections-on-the-green-revolutions-narrative-and-myths/.

19. Bill Gates, "2009 World Food Prize Symposium," prepared remarks, October 15, 2009, https://www.gatesfoundation.org/Media-Center/Speeches/2009/10/Bill-Gates-2009-World-Food-Prize-Symposium.

20. Harwood, "Could the Adverse Consequences," 511.

21. Baranski and Ollenburger, "How to Improve the Social Benefits."

22. M. MacNeil, "'Better, Faster, Equitable, Sustainable'—Wheat Research Community Partners Join to Kick off New Breeding Project," CIMMYT, July 29, 2020, https://www.cimmyt.org/news/better-faster-equitable-sustainable-wheat-research-community-partners-join-to-kick-off-new-breeding-project.

23. MacNeil, "'Better, Faster, Equitable, Sustainable.'"

24. The CGIAR does recognize these different pathways, and the CGIAR's program on Climate Change, Agriculture and Food Security was more interdisciplinary and social science–focused than other programs.

25. Fischer, "Why New Crop Technology," 1185–86.

26. Ollenburger et al., "Are Farmers Searching," 2.

27. Ollenburger et al., "Are Farmers Searching," 16.

28. Harris and Orr, "Is Rainfed Agriculture."

29. National Resources Institute, University of Greenwich, *End of Program Evaluation for AGRA*, 94.

30. Ekboir, *CGIAR at a Crossroads*, 4.

31. Bill Gates, "Bill Gates on Adapting to a Warmer World," *Financial Post*, February 11, 2021, https://financialpost.com/pmn/business-pmn/bill-gates-on-adapting-to-a-warmer-world.

32. Schurman, "Micro(soft) Managing a 'Green Revolution,'" 189.

33. Brooks et al. *Environmental Change*, 27.

34. Schurman, "Micro(soft) Managing a 'Green Revolution,'" 181.

35. See Baranski and Ollenburger, "How to Improve the Social Benefits," for more context.

36. Brooks et al., *Environmental Change*.

37. Brooks et al., *Environmental Change*, 27.

38. Mastenbroek et al., *Community Based Risk Spectrum Analysis*, 2.

39. Fisher and Carr, "Influence of Gendered Roles"; Makate, Makate, and Mango, "Wealth-Related Inequalities."

40. Atlin, Cairns, and Das, "Rapid Breeding."

41. See National Resources Institute, University of Greenwich, *End of Program Evaluation for AGRA*.

42. Schurman, "Micro(soft) Managing a 'Green Revolution,'" 187.

43. Harwood, *Europe's Green Revolution and Its Successors*, 162.

44. Schurman, "Micro(soft) Managing a 'Green Revolution,'" 187.

45. Harwood, *Europe's Green Revolution and Its Successors*.

46. Ollenburger et al., "Are Farmers Searching," 20.

BIBLIOGRAPHY

Abel, Martin. E. "Agriculture in India in the 1970s." *Economic and Political Weekly* 5, no. 13 (March 28, 1970): A5–A14.

Abrol, Dinesh. "American Involvement in Indian Agricultural Research." *Social Scientist* 11, no. 10 (1983): 8–26.

Acharya, Nilachala, and Subrat Das. "Revitalising Agriculture in Eastern India: Investment and Policy Priorities." *IDS Bulletin* 43, no. 1 (2012): 104–12.

Adas, Michael. *Machines as the Measure of Men: Science, Technology, and Ideologies of Western Dominance.* Ithaca, NY: Cornell University Press, 1989.

Agrawal, R. K. "Development of Improved Varieties." In *Twenty-Five Years of Coordinated Wheat Research, 1961–1986,* edited by J. P. Tandon and A. P. Sethi, 34–93. Delhi: Wheat Project Directorate, Indian Agricultural Research Institute, 1986.

Ahlberg, Kristin L. "'Machiavelli with a Heart': The Johnson Administration's Food for Peace Program in India, 1965–1966." *Diplomatic History* 31, no. 4 (2007): 665–701.

Anderson, R. G. *Potentials for Improving Production Efficiency of Cereals.* Mexico City, Mexico: CIMMYT, 1972.

Anderson, R. G., ed. *Wheat, Triticale and Barley Seminar: Proceedings.* El Batán, Mexico: CIMMYT, 1973.

Anderson, Robert S. "Cultivating Science as Cultural Policy: A Contrast of Agricultural and Nuclear Science in India." *Pacific Affairs* 56 (1983): 38–50.

Anderson, Robert S., Paul R. Brass, Edwin Levy, and Barrie. M. Morrison, eds. *Science, Politics, and the Agricultural Revolution in Asia.* Boulder, CO: Westview, 1982.

Anderson, Robert S., Edwin Levy, and Barrie M. Morrison. *Rice Science and Development Politics: Research Strategies and IRRI's Technologies Confront Asian Diversity (1950–1980).* Oxford: Clarendon, 1991.

Annicchiarico, Paolo. *Genotype × Environment Interactions: Challenges and Opportunities for Plant Breeding and Cultivar Recommendations.* Rome: Food and Agriculture Organization, 2002. http://www.fao.org/docrep/005/y4391e/y4391e00.htm.

Arce, Alberto. "Bureaucratic Conflict and Public Policy: Rainfed Agriculture in Mexico." *Boletín de Estudios Latinoamericanos y del Caribe* 42 (1987): 3–24.

Aronova, Elena, Karen S. Baker, and Naomi Oreskes. "Big Science and Big Data in Biology: From the International Geophysical Year through the International Biological Program to the Long Term Ecological Research (LTER) Network, 1957–Present." *Historical Studies in the Natural Sciences* 40, no. 2 (2010): 183–224.

Arora, Saurabh, Barbara Van Dyck, Divya Sharma, and Andy Stirling. "Control, Care, and Conviviality in the Politics of Technology for Sustainability." *Sustainability: Science, Practice and Policy* 17, no. 2 (2020): 247–62.

Athwal, D. S., and Gian Singh. "Variability in *Kangni*—1. Adaptation and Genotypic and Phenotypic Variability in Four Environments." *Indian Journal of Genetics and Plant Breeding* 26, no. 2 (1966): 142–52.

Atlin, Gary N., Jill E. Cairns, and Biswanath Das. "Rapid Breeding and Varietal Replacement Are Critical to Adaptation of Cropping Systems in the Developing World to Climate Change." *Global Food Security* 12 (2017): 31–37.

Babu, Suresh Chandra, P. K. Joshi, Claire J. Glendenning, Kwadwo Asenso-Okyere, and Rasheed Sulaiman. "The State of Agricultural Extension Reforms in India: Strategic Priorities and Policy Options." *Agricultural Economics Research Review* 26 (2013): 159–72.

Balaguru, T., K. Venkateswarlu, and M. Rajagopalan. "Implementation of World Bank Aided National Agricultural Research Project in India: A Case Study." *Agricultural Administration and Extension* 29, no. 2 (1988): 135–47.

Bänziger, Marianne, and Mark Cooper. "Breeding for Low Input Conditions and Consequences for Participatory Plant Breeding Examples from Tropical Maize and Wheat." *Euphytica* 122 (2001): 503–19.

Baranski, Marci. "The Wide Adaptation of Green Revolution Wheat." PhD diss., Arizona State University, 2015.

Baranski, Marci, and Mary Ollenburger. "How to Improve the Social Benefits of Agricultural Research." *Issues in Science and Technology* 36, no. 3 (2020): 47–53.

Barnes, Jessica. *Transnational Flows of Expertise and Seed in the Making of*

Egypt's Wheat. Rockefeller Archive Center Research Reports, October 7, 2019. https://rockarch.issuelab.org/resource/transnational-flows-of-exper tise-and-seed-in-the-making-of-egypt-s-wheat.html.

Bashford, Alison. "Population, Geopolitics, and International Organizations in the Mid-twentieth Century." *Journal of World History* 19, no. 3 (2008): 327–47.

Basu, Santonu. "Government Success, Failure of the Market: A Case Study of Rural India." *International Review of Applied Economics* 23 (2009): 485–501.

Baum, Warren C. *Partners against Hunger: Consultative Group on International Agricultural Research.* Washington, DC: World Bank, 1986.

Bellon, Mauricio R., and Jean Risopoulos. "Small-Scale Farmers Expand the Benefits of Improved Maize Germplasm: A Case Study from Chiapas, Mex- ico." *World Development* 29, no. 5 (2001): 799–811.

Bhullar, G. S., K. S. Gill, and A. S. Khehra. "Stability Analysis over Various Filial Generations in Bread Wheat." *Theoretical and Applied Genetics* 51 (1977): 41–44.

Bidinger, F. R., V. Mahalakshmi, and G. D. P. Rao. "Assessment of Drought Re- sistance in Pearl Millet [*Pennisetum americanum* (L.) Leeke]. I. Factors Af- fecting Yields under Stress." *Australian Journal of Agricultural Research* 38, no. 1 (1987): 37–48.

Biggs, Stephen D. "A Multiple Source of Innovation Model of Agricultural Re- search and Technology Promotion." *World Development* 18, no. 11 (1990): 1481–99.

Borlaug, Norman E. "Breeding Wheat for High Yield, Wide Adaptation, and Disease Resistance." In *Rice Breeding*, edited by International Rice Research Institute, 581–92. Los Baños, Philippines: IRRI, 1972.

Borlaug, Norman E. "A Cereal Breeder and Ex-Forester's Evaluation of the Prog- ress and Problems Involved in Breeding Rust Resistant Forest Trees: Mod- erator's Summary." In *Biology of Rust Resistance in Forest Trees: Proceedings of a NATO-IUFRO Advanced Study Institute, August 17–24, 1969*, edited by United States Department of Agriculture, 615–42. Washington, DC: US Forest Service, 1972.

Borlaug, Norman E. "The Green Revolution: The Role of CIMMYT and What Lies Ahead." In *Trade and Development: Proceedings of the Winter 1986 Meeting of the International Agricultural Trade Research Consortium (Tex- coco, Mexico, December 1986)*, edited by Mathew D. Shane, 113–27. Wash- ington DC: US Department of Agriculture, Economic Research Service, 1988.

Borlaug, Norman E. "Wheat Breeding and Its Impact on World Food Supply." In *Proceedings of the Third International Wheat Genetics Symposium*, edited by Keith W. Finlay and Kenneth W. Shepherd, 1–36. Boston: Butterworth, 1968.

Borlaug, Norman E., Jacobo Ortega, and Alfredo García. *Preliminary Report of the Results of the First Cooperative Inter-American Spring Wheat Yield Nursery Grown during 1960–61.* El Batán, Mexico: CIMMYT, 1964.

Borlaug, Norman E., Jacobo Ortega, and Ricardo Rodriguez. *Preliminary Report on the Result of the Second Cooperative Near East–American Spring Wheat Yield Nursery, 1962–63.* Miscellaneous Report No. 5. El Batán, Mexico: CIMMYT, 1964.

Bozeman, Barry, and Daniel Sarewitz. "Public Values and Public Failure in US Science Policy." *Science and Public Policy* 32, no. 2 (2005): 119–36.

Brancourt-Hulmel, M., E. Heumez, P. Pluchard, D. Beghin, C. Depatureaux, A. Giraud, and J. Le Gouis. "Indirect versus Direct Selection of Winter Wheat for Low-Input or High-Input Levels." *Crop Science* 45, no. 4 (2005): 1427–31.

Braun, Hans-Joachim, Sanjaya Rajaram, and Maarten van Ginkel. "CIMMYT's Approach to Breeding for Wide Adaptation." *Euphytica* 92 (1996): 175–83.

Braun, H.-J., N. Zencirci, F. Altay, A. Atli, M. Avci, V. Eser, M. Kambertay, and T. S. Payne. "Turkish Wheat Pool." In *The World Wheat Book: A History of Wheat Breeding,* edited by Alain P. Bonjean and William J. Angus, 851–79. London: Lavoisier, 2001.

Breth, Steven A. "Turkey's Wheat Research and Training Project." *CIMMYT Today* 6 (1977): 1–19.

Brooks, Sally, John Thompson, Hannington Odame, Betty Kibaara, Serah Nderitu, Francis Karin, and Erik Millstone. *Environmental Change and Maize Innovation in Kenya: Exploring Pathways in and out of Maize.* STEPS working paper 36. Brighton, UK: STEPS Centre, 2009.

Busch, Lawrence. *Universities for Development: Report of the Joint Indo-U.S. Impact Evaluation of the Indian Agricultural Universities.* Washington, DC: US Agency for International Development, 1988.

Byerlee, Derek. *The Birth of CIMMYT: Pioneering the Idea and Ideals of International Agricultural Research.* El Batán, Mexico: CIMMYT, 2016.

Cabral, Lídia, Poonam Pandey, and Xiuli Xu. "Epic Narratives of the Green Revolution in Brazil, China, and India." *Agriculture and Human Values* 39 (2021): 249–67.

Calhoun, D. S., G. Gebeyehu, A. Miranda, S. Rajaram, and M. Van Ginkel. "Choosing Evaluation Environments to Increase Wheat Grain Yield under Drought Conditions." *Crop Science* 34, no. 3 (1994): 673–78.

Cash, David W., Jonathan C. Borck, and Anthony G. Patt. "Countering the Loading-Dock Approach to Linking Science and Decision Making: Comparative Analysis of El Niño/Southern Oscillation (ENSO) Forecasting Systems." *Science, Technology and Human Values* 31, no. 4 (2006): 465–94.

Ceccarelli, Salvatore. "Adaptation to Low/High Input Cultivation." *Euphytica* 92 (1996): 203–14.

Ceccarelli, S. "Wide Adaptation: How Wide?" *Euphytica* 40 (1989): 197–205.

Ceccarelli, S. "Yield Potential and Drought Tolerance of Segregating Populations of Barley in Contrasting Environments." *Euphytica* 36 (1987): 265–73.

Ceccarelli, S., W. Erskine, J. Hamblin, and S. Grando. "Genotype by Environment Interaction and International Breeding Programmes." *Experimental Agriculture* 30, no. 2 (1994): 177–87.

CGIAR-IEA. *Evaluation of CGIAR Research Program on Wheat*. Rome: Independent Evaluation Arrangement of CGIAR, 2015.

CGIAR Technical Advisory Committee. *TAC Statement on the Division of Responsibility for Durum Wheat and Barley between CIMMYT and ICARDA*. Washington, DC: CGIAR, 1983.

Chakravarti, A. K. "Green Revolution in India." *Annals of the Association of American Geographers* 63, no. 3 (1973): 319–30.

Chambers, Robert. *Paradigm Shifts and the Practice of Participatory Research and Development*. Institute of Development Studies Working Paper 2. Brighton: Institute of Development Studies, 1994.

Chandra, S. "Variability in Gram." *Indian Journal of Genetics and Plant Breeding* 28, no. 2 (1968): 205–10.

Chatterjee, S., and D. Kapur. "Understanding Price Variation in Agricultural Commodities in India: MSP, Government Procurement, and Agriculture Markets." NCAER India Policy Forum, New Delhi, July 12–13, 2016.

CIMMYT. *Annual Report, 1972*. Mexico City: CIMMYT, 1972.

CIMMYT. *CIMMYT Report, 1967–68*. Mexico City: CIMMYT, 1968.

CIMMYT. *CIMMYT Report, 1968–69*. Mexico City: CIMMYT, 1969.

CIMMYT. *CIMMYT Report on Wheat Improvement: 1980*. Mexico City: CIMMYT, 1980.

CIMMYT. *Enduring Designs for Change: An Account of CIMMYT's Research, Its Impact, and Its Future Directions*. Mexico City: CIMMYT, 1992.

CIMMYT. *The Puebla Project: Seven Years of Experience: 1967–1973*. El Batán, Mexico: CIMMYT, 1974.

CIMMYT. *Report on Wheat Improvement*. El Batán, Mexico: CIMMYT, 1984.

CIMMYT and ICARDA. *CRP 3.1: Wheat: Global Alliance for Improving Food Security and the Livelihoods of the Resource-Poor in the Developing World*. El Batán, Mexico: CIMMYT, 2011.

Cleaver, Harry M., Jr. "The Contradictions of the Green Revolution." *American Economic Review* 62, no. 1/2 (1972): 177–86.

Cleveland, David A. "Is Plant Breeding Science Objective Truth or Social Construction? The Case of Yield Stability." *Agriculture and Human Values* 18 (2001): 251–70.

Colombian Agricultural Program. *Director's Annual Report: May 1, 1955–April 30, 1956*. New York: Rockefeller Foundation, 1956.

Colombian Agricultural Program. *Director's Annual Report: May 1, 1956–April 30, 1957*. New York: Rockefeller Foundation, 1957.

Connelly, Matthew. *Fatal Misconception: The Struggle to Control World Population*. Cambridge, MA: Belknap Press of Harvard University Press, 2008.

Cooper, M., R. E. Stucker, I. H. DeLacy, and B. D. Harch. "Wheat Breeding Nurseries, Target Environments, and Indirect Selection for Grain Yield." *Crop Science* 37, no. 4 (1997): 1168–76.

Cooper, M., D. R. Woodruff, R. L. Eisemann, P. S. Brennan, and I. H. DeLacy. "A Selection Strategy to Accommodate Genotype-by-Environment Interaction for Grain Yield of Wheat: Managed-Environments for Selection among Genotypes." *Theoretical and Applied Genetics* 90 (1995): 492–502.

Cotter, Joseph. *Troubled Harvest: Agronomy and Revolution in Mexico, 1880–2002*. Westport, CT: Praeger, 2003.

Cowen, Michael, and Robert W. Shenton. *Doctrines of Development*. Abingdon, UK: Routledge, 1996.

Cullather, Nick. *The Hungry World: America's Cold War Battle against Poverty in Asia*. Cambridge, MA: Harvard University Press, 2010.

Cullather, Nick. "Miracles of Modernization: The Green Revolution and the Apotheosis of Technology." *Diplomatic History* 28, no. 2 (2004): 227–54.

Curry, Helen Anne. "From Working Collections to the World Germplasm Project: Agricultural Modernization and Genetic Conservation at the Rockefeller Foundation." *History and Philosophy of the Life Sciences* 39, no. 2 (2017): article 5.

Dahlberg, Kenneth A. *Beyond the Green Revolution: The Ecology and Politics of Global Agricultural Development*. New York: Plenum, 1979.

Dahlberg, Kenneth A. "Ethical and Value Issues in International Agricultural Research." *Agriculture and Human Values* 5 (1988): 101–11.

Das, P. K. and H. K. Jain. "Studies on Adaptation in Wheat. II. Unilocation Testing for Phenotypic Stability and Varietal Adaptability." *Indian Journal of Genetics and Plant Breeding* 31, no. 1 (1971): 77–85.

Das, Raju J. "The Green Revolution and Poverty: A Theoretical and Empirical Examination of the Relation between Technology and Society." *Geoforum* 33, no. 1 (2002): 55–72.

Davis, Mike. *Late Victorian Holocausts: El Niño Famines and the Making of the Third World*. New York: Verso, 2002.

Dawson, Julie C., Kevin M. Murphy, and Stephen S. Jones. "Decentralized Selection and Participatory Approaches in Plant Breeding for Low-Input Systems." *Euphytica* 160 (2008): 143–54.

Desai, Bhupat, Errol D'Souza, John W. Mellor, Vijay P. Sharma, and Prabhakar Tamboli. "Agricultural Policy Strategy, Instruments and Implementation: A Review and the Road Ahead." *Economic and Political Weekly* 46, no. 53 (December 31, 2011): 42–50.

Desai, Gunvant M. "Fertiliser Use in India: The Next Stage in Policy." *Indian Journal of Agricultural Economics* 41, no. 3 (1986): 248–70.

Desai, Sonalde, Lawrence J. Haddad, Deepta Chopra, and Amit Thorat, eds. *Undernutrition and Public Policy in India: Investing in the Future*. Abingdon, UK: Routledge, 2016.

Descheemaeker, Katrien, Esther Ronner, Mary Ollenburger, Angelinus C. Franke, Charlotte J. Klapwijk, Gatien N. Falconnier, Jannike Wichern, and Kenneth E. Giller. "Which Options Fit Best? Operationalizing the Socio-ecological Niche Concept." *Experimental Agriculture* 55, no. S1 (2019): 169–90.

Dhanagare, D. N. "Green Revolution and Social Inequalities in Rural India." *Economic and Political Weekly* 22, no. 19/21 (1987): AN137–44.

Douthwaite, Boru. *Enabling Innovation: A Practical Guide to Understanding and Fostering Technological Change*. London: Zed Books, 2002.

Douthwaite, B., J. D. H. Keatinge, and J. R. Park. "Why Promising Technologies Fail: The Neglected Role of User Innovation during Adoption." *Research Policy* 30, no. 5 (2001): 819–36.

Duncan, Andy. "Wheat Dreams." *Oregon's Agricultural Progress* 41, no. 3 (1995): 8–15.

Dworkin, Susan. *The Viking in the Wheat Field: A Scientist's Struggle to Preserve the World's Harvest*. New York: Walker, 2009.

Easter, K. William, S. Bisaliah, and John O. Dunbar. "After Twenty-Five Years of Institution Building, the State Agricultural Universities in India Face New Challenges." *American Journal of Agricultural Economics* 71, no. 5 (1989): 1200–1205.

Ekboir, J. *The CGIAR at a Crossroads: Assessing the Role of International Agricultural Research in Poverty Alleviation from an Innovation Systems Perspective*. ILAC Working Paper 9. Silverwater, Australia: Institutional Learning and Change, 2009.

Eliazar Nelson, Ann Raeboline Lincy, Kavitha Ravichandran, and Usha Antony. "The Impact of the Green Revolution on Indigenous Crops of India." *Journal of Ethnic Foods* 6 (2019): article 8.

"Ending Hunger: Science Must Stop Neglecting Smallholder Farmers." *Nature* 586 (2020): 336.

Evans, L. T. *Crop Evolution, Adaptation and Yield*. Cambridge: Cambridge University Press, 1996.

Farmer, B. H. "The 'Green Revolution' in South Asian Ricefields: Environment and Production." *Journal of Development Studies* 15, no. 4 (1979): 304–19.

Finlay, Keith W. "The Significance of Adaptation in Wheat Breeding." In *Proceedings of the Third International Wheat Genetics Symposium*, edited by Keith W. Finlay and Kenneth William. Shepherd, 403–9. Boston: Butterworth, 1968.

Finlay, K. W., and G. N. Wilkinson. "The Analysis of Adaptation in a Plant-Breeding Programme." *Australian Journal of Agricultural Research* 14, no. 6 (1963): 742–54.

Fischer, Klara. "Why New Crop Technology Is Not Scale-Neutral—A Critique of the Expectations for a Crop-Based African Green Revolution." *Research Policy* 45, no. 6 (2016): 1185–94.

Fischer, Klara, and Flora Hajdu. "Does Raising Maize Yields Lead to Poverty Reduction? A Case Study of the Massive Food Production Programme in South Africa." *Land Use Policy* 46 (2015): 304–13.

Fischer, R. A., and R. Maurer. "Drought Resistance in Spring Wheat Cultivars. I. Grain Yield Responses." *Australian Journal of Agricultural Research* 29, no. 5 (1978): 897–912.

Fisher, Monica, and Edward R. Carr. "The Influence of Gendered Roles and Responsibilities on the Adoption of Technologies That Mitigate Drought Risk: The Case of Drought-Tolerant Maize Seed in Eastern Uganda." *Global Environmental Change* 35 (2015): 82–92.

Fitzgerald, Deborah. "Exporting American Agriculture: The Rockefeller Foundation in Mexico, 1943–53." *Social Studies of Science* 16, no. 3 (1986): 457–83.

Food and Agriculture Organization of the United Nations. *Fertilizer Use by Crop in India*. Rome: FAO, 2005.

Food and Agriculture Organization of the United Nations. *A Study on the Response of Wheat to Fertilizers*. FAO Soils Bulletin 12. Rome, 1971.

Frankel, Francine R. *India's Green Revolution: Economic Gains and Political Costs*. Princeton, NJ: Princeton University Press, 1971.

Frankel, Francine R. "India's New Strategy of Agricultural Development: Political Costs of Agrarian Modernization." *Journal of Asian Studies* 28, no. 4 (1969): 693–710.

Frankel, O. H. "Adaptability of Crops." *New Scientist* 31 (1966): 144–45.

Freebairn, Donald K. "Did the Green Revolution Concentrate Incomes? A Quantitative Study of Research Reports." *World Development* 23, no. 2 (1995): 265–79.

Fuglie, Keith, Madhur Gautam, Aparajita Goyal, and William F. Maloney. *Harvesting Prosperity: Technology and Productivity Growth in Agriculture*. Washington, DC: World Bank, 2020.

Gangopadhyaya, M., and R. P. Sarker. "Influence of Rainfall Distribution on the Yield of Wheat Crop." *Agricultural Meteorology* 2, no. 5 (1965): 331–50.

Glaeser, Bernhard, ed. *The Green Revolution Revisited: Critique and Alternatives*. Abingdon, UK: Routledge, 2011.

Glover, Dominic, and Nigel Poole. "Principles of Innovation to Build Nutrition-Sensitive Food Systems in South Asia." *Food Policy* 82 (2019): 63–73.

Gourdji, Sharon M., Ky L. Mathews, Matthew Reynolds, José Crossa, and David B. Lobell. "An Assessment of Wheat Yield Sensitivity and Breeding Gains in Hot Environments." *Proceedings of the Royal Society of London B: Biological Sciences* 280, no. 1752 (2013): article 20122190.

Government of India. *Implementation of the Programme of 'Bringing the Green*

Revolution in Eastern India (BGREI)'- A Sub Scheme of RVKY during the Year 2014–15. New Delhi: Department of Agriculture and Cooperation Ministry of Agriculture, 2014.

Government of India. *Intensive Agricultural Districts Programme Report of the First Central Training Course for the Seed Development Officers*. New Delhi: Ministry of Food and Agriculture, Directorate of Extension, 1961.

Government of India. *National Workshop on Decentralized Seed Systems for Climate Resilient Agriculture in Rainfed Areas*. New Delhi: National Institute of Agricultural Extension Management, Ministry of Agriculture, Government of India, 2019.

Government of India. *Report of the National Commission on Agriculture 1976: Part I*. New Delhi: Ministry of Agriculture and Irrigation, 1976.

Government of India. *Twelfth Five Year Plan (2012–2017)*. New Delhi: Planning Commission, Government of India, 2013.

Government of India, Directorate of Economics and Statistics. *Agricultural Statistics at a Glance 2015*. Government of India Controller of Publication, 2017. http://eands.dacnet.nic.in/PDF/Agricultural_Statistics_At_Glance-2015 .pdf

Govind, Nalini. *Regional Perspectives in Agricultural Development: A Case Study of Wheat and Rice in Selected Regions of India*. New Delhi: Concept, 1986.

Greulach, Victor A. "Photoperiodism: The Remarkable Influence of Length-of-Day on Plant Processes." *School Science and Mathematics* 33, no. 7 (1933): 707–20.

Griffin, Keith. *The Political Economy of Agrarian Change: An Essay on the Green Revolution*. Cambridge, MA: Harvard University Press, 1974.

Guha, Ramachandra. *India after Gandhi: The History of the World's Largest Democracy*. New York: Macmillan, 2007.

Gupta, V. P., D. S. Virk, and D. R. Satija. "Quantitative Genetic Analysis in Wheat." In *Genetics and Wheat Improvement: Proceedings of the 2nd National Seminar on Genetics and Wheat Improvement*, edited by Akshey K. Gupta, 75–90. New Delhi: Oxford and IBH, 1980.

Hall, Andrew, Norman Clark, Rasheed Sulaiman, M. V. K. Sivamohan, and B. Yoganand. "New Agendas for Agricultural Research in Developing Countries: Policy Analysis and Institutional Implications." *Knowledge, Technology and Policy* 13 (2000): 70–91.

Hall, Andrew, Rasheed Sulaiman, Norman Clark, M. V. K. Sivamohan, and B. Yoganand. "Public-Private Sector Interaction in the Indian Agricultural Research System: An Innovation Systems Perspective on Institutional Reform." In *Agricultural Research Policy in an Era of Privatization*, edited by Derek Byerlee and Ruben G. Echeverría, 155–76. Wallingford, UK: CABI, 2002.

Harris, David, and Alastair Orr. "Is Rainfed Agriculture Really a Pathway from Poverty?" *Agricultural Systems* 123 (2014): 84–96.

Harwood, Jonathan. "Another Green Revolution? On the Perils of 'Extracting Lessons' from History." *Development* 61 (2018): 43–53.

Harwood, Jonathan. "Could the Adverse Consequences of the Green Revolution Have Been Foreseen? How Experts Responded to Unwelcome Evidence." *Agroecology and Sustainable Food Systems* 44, no. 4 (2020): 509–35.

Harwood, Jonathan. *Europe's Green Revolution and Its Successors: The Rise and Fall of Peasant-Friendly Plant Breeding.* Abingdon, UK: Routledge, 2012.

Harwood, Jonathan. "Global Visions vs. Local Complexity: Experts Wrestle with the Problem of Development." In *Rice: Global Networks and New Histories,* edited by Francesca Bray, Peter A. Coclanis, Edda L. Fields-Black, and Dagmar Schaefer, 41–55. Cambridge: Cambridge University Press, 2015.

Harwood, Jonathan. "Peasant Friendly Plant Breeding and the Early Years of the Green Revolution in Mexico." *Agricultural History* 83, no. 3 (2009): 384–410.

Harwood, Jonathan. "Was the Green Revolution Intended to Maximise Food Production?" *International Journal of Agricultural Sustainability* 17, no. 4 (2019): 312–25.

Hazell, Peter B., ed. *Summary Proceedings of a Workshop on Cereal Yield Variability.* Washington, DC: International Food Policy Research Institute, 1986.

Hesser, Leon. *The Man Who Fed the World: Nobel Peace Prize Laureate Norman Borlaug and His Battle to End World Hunger.* Dallas: Durban House, 2006.

Hill, Julian, Heiko C. Becker, and Peter M. A. Tigerstedt. *Quantitative and Ecological Aspects of Plant Breeding.* London: Chapman and Hall, 1998.

Horner, T. W., and K. J. Frey. "Methods for Determining Natural Areas for Oat Varietal Recommendations." *Agronomy Journal* 49, no. 6 (1957): 313–15.

Howard, Albert, and Gabrielle L. C. Howard. *Wheat in India: Its Production, Varieties, and Improvement.* Calcutta: Imperial Department of Agriculture in India by Thacker, Spink, 1909.

Hughes, Thomas P. "Technological Momentum." In *Does Technology Drive History? The Dilemma of Technological Determinism,* edited by Merritt Roe Smith and Leo Marx, 101–13. Cambridge: MIT Press, 1994.

Hurd, E. A. "A Method of Breeding for Yield of Wheat in Semi-arid Climates." *Euphytica* 18 (1969): 217–26.

Hurd, E. A. "Phenotype and Drought Tolerance in Wheat." In *Developments in Agricultural and Managed Forest Ecology,* vol. 1, edited by J. Stone, 39–55. Amsterdam: Elsevier, 1975.

International Food Policy Research Institute. *Global Food Policy Report, 2014–2015.* Washington, DC: International Food Policy Research Institute, 2015.

International Rice Research Institute. *IR8 and Beyond.* Los Baños, Philippines: IRRI, 1977.

Jefferson, Osmat A. "How Are Accountability Standards Implemented in International Agricultural Centers?" *ILSA Journal of International and Comparative Law* 16 (2010): 457–67.

Jennings, Bruce H. *Foundations of International Agricultural Research: Science and Politics in Mexican Agriculture.* Boulder, CO: Westview, 1988.

Jones, David M. "The Green Revolution in Latin America: Success or Failure?" In "International Aspects of Development in Latin America: Geographic Aspects," vol. 6 of *Publication Series (Conference of Latin Americanist Geographers)* (1977): 55–63.

Joshi, A. K., B. Mishra, R. Chatrath, G. Ortiz Ferrara, and Ravi P. Singh. "Wheat Improvement in India: Present Status, Emerging Challenges and Future Prospects." *Euphytica* 157 (2007): 431–46.

Joshi, K. D., A. M. Musa, C. Johansen, S. Gyawali, D. Harris, and J.R. Witcombe. "Highly Client-Oriented Breeding, Using Local Preferences and Selection, Produces Widely Adapted Rice Varieties." *Field Crops Research* 100, no. 1 (2007): 107–16.

Kalirajan, K., and R. T. Shand. "Location Specific Research: Rice Technology in India." *Land Economics* 58, no. 4 (1982): 537–46.

Kanwar, J. S. "From Protective to Productive Irrigation." *Economic and Political Weekly* 4, no. 13 (March 29, 1969): A21–A26.

Kanwar, J. S. "Research for Effective Use of Land and Water Resources." In *National Agricultural Research Systems in Asia*, edited by Albert H. Moseman, 214–25. Washington, DC: Agricultural Development Council, 1971.

Kashyap, S. P., and Niti Mathur. "Ongoing Changes in Policy Environment and Farm Sector: Role of Agro-climatic Regional Planning Approach." *Economic and Political Weekly* 34, no. 26 (June 26, 1999): A105–A112.

Katyal, J. C., and Mrutyunjaya. *CGIAR Effectiveness—A NARS Perspective from India.* Operations Evaluation Department Working Paper Series. Washington, DC: World Bank Group, 2003.

Keser, Mesut, Nurberdy Gummadov, Beyhan Akin, Savas Belen, Zafer Mert, Seyfi Taner, Ali Topal, Selami Yazar, Alexey Morgounov, Ram Chandra Sharma, and Fatih Ozdemir. "Genetic Gains in Wheat in Turkey: Winter Wheat for Dryland Conditions." *Crop Journal* 5 (2017): 533–40.

Klatt, A. "Breeding for Yield Potential, Stability, and Adaptation." In *Wheat, Triticale, and Barley Seminar: Proceedings*, edited by R. G. Anderson, 104–13. El Batán, Mexico: CIMMYT, 1973.

Klerkx, Laurens, and Cees Leeuwis, "Matching Demand and Supply in the Agricultural Knowledge Infrastructure: Experiences with Innovation Intermediaries." *Food Policy* 33, no. 3 (2008): 260–76.

Kloppenburg, Jack Ralph. *First the Seed: The Political Economy of Plant Biotechnology.* Madison: University of Wisconsin Press, 2005.

Knight, Henry. *Food Administration in India, 1939–47.* Palo Alto, CA: Stanford University Press, 1954.

Kohli, S. P. "Towards Further Rationalization in Plant Breeding." *Indian Journal of Genetics and Plant Breeding* 29 (1969): 24–29.

Krishna, Daya. *The New Agricultural Strategy: The Vehicle of Green Revolution in India*. New Delhi: New Heights, 1971.

Krishna, V. V., S. Aravindakshan, A. Chowdhury, and B. Rudra. *Farmer Access and Differential Impacts of Zero Tillage Technology in the Subsistence Wheat Farming Systems of West Bengal, India*. Socioeconomics Working Paper 7. Mexico City: CIMMYT, 2012.

Krishna, V. V., D. J. Spielman, and P. C. Veettil. "Exploring the Supply and Demand Factors of Varietal Turnover in Indian Wheat." *Journal of Agricultural Science* 154 (2016): 258–72.

Krishna, V. V., D. J. Spielman, P. C. Veettil, and S. Ghimire. *An Empirical Examination of the Dynamics of Varietal Turnover in Indian Wheat*. IFPRI Discussion Paper 01336. Washington, DC: International Food Policy Research Institute, 2014.

Krishna, Vijesh V., Yigezu A. Yigezu, Aziz Karimov, and Olaf Erenstein. "Assessing Technological Change in Agri-food Systems of the Global South: A Review of Adoption-Impact Studies in Wheat." *Outlook on Agriculture* 49, no. 2 (2020): 89–98.

Kronstad, Warren E. *Global Report on the Introduction of Semi Dwarf Wheats to Less Developed Countries*. Washington, DC: US Agency for International Development, 1969.

Krull, C. F., I. Narvaez, N. E. Borlaug, J. Ortega, G. Vasquez, R. Rodriguez, and C. Meza. *Results of the Fourth Inter-American Spring Wheat Yield Nursery, 1963–1964*. CIMMYT Research Bulletin no. 7, 1967.

Krull, C. F., I. Narvaez, N. E. Borlaug, J. Ortega, G. Vasquez, R. Rodriguez, and C. Meza. *Results of the Third Near East–American Spring Wheat Yield Nursery, 1963–65*. CIMMYT Research Bulletin no. 5, November 1966.

Krull, Charles F., Norman E. Borlaug, Carlos Meza, and I. Narváez. *Results of the First International Spring Wheat Yield Nursery, 1964–1965*. CIMMYT Research Bulletin no. 8. Mexico City: CIMMYT, 1968.

Kumar, M. Dinesh. *Water Policy Science and Politics: An Indian Perspective*. Amsterdam: Elsevier, 2018.

Kumar, Prakash. "'Modernization' and Agrarian Development in India, 1912–52." *Journal of Asian Studies* 79, no. 3 (2020): 633–58.

Kumar, Richa. *Rethinking Revolutions: Soyabean, Choupals, and the Changing Countryside in Central India*. Oxford: Oxford University Press, 2016.

Ladejinsky, Wolf. "The Green Revolution in Punjab: A Field Trip." *Economic and Political Weekly* 4, no. 26 (June 28, 1969): A73–A82.

Ladejinsky, Wolf. "Ironies of India's Green Revolution." *Foreign Affairs* 48 (1970): 758–68.

Laing, D. R., and R. A. Fischer. "Adaptation of Semidwarf Wheat Cultivars to Rainfed Conditions." *Euphytica* 26 (1977): 129–39.

Lancaster, Carol. *Foreign Aid: Diplomacy, Development, Domestic Politics.* Chicago: University of Chicago Press, 2008.

Lantican, M.A., H.-J. Braun, T. S. Payne, R. P. Singh, S. Sonder, M. Baum, M. van Ginkel, and O. Erenstein. *Impacts of International Wheat Improvement Research, 1994–2014.* Mexico City: CIMMYT, 2016.

Lantican, M. A., H. J. Dubin, and M. L. Morris. *Impacts of International Wheat Breeding Research in the Developing World, 1988–2002.* Mexico City: CIMMYT, 2005.

Latham, Michael E. *The Right Kind of Revolution: Modernization, Development, and U.S. Foreign Policy from the Cold War to the Present.* Ithaca, NY: Cornell University Press, 2011.

Leach, Melissa, Ian Scoones, and Andy Stirling. *Dynamic Sustainabilities: Technology, Environment, Social Justice.* London: Earthscan, 2010.

Lele, Uma, and Arthur A. Goldsmith. "The Development of National Agricultural Research Capacity: India's Experience with the Rockefeller Foundation and Its Significance for Africa." *Economic Development and Cultural Change* 37, no. 2 (1989): 305–43.

Lewontin, Stephen. "The Green Revolution and the Politics of Agricultural Development in Mexico since 1940." PhD diss., University of Chicago, 1983.

Loegering, W. Q., and N. E. Borlaug. *Contribution of the International Spring Wheat Rust Nursery to Human Progress and International Good Will.* Washington DC: US Department of Agriculture, Agricultural Research Service, 1963.

Luthra, O. P., and R. K. Singh. "A Comparison of Different Stability Models in Wheat." *Theoretical and Applied Genetics* 45, no. 4 (1974): 143–49.

MacDonald, A. M., H. C. Bonsor, K. M. Ahmed, W. G. Burgess, M. Basharat, R. C. Calow, A. Dixit, et al. "Groundwater Quality and Depletion in the Indo-Gangetic Basin Mapped from In Situ Observations." *Nature Geoscience* 9 (2016): 762–66.

Majumdar, Kaushik, Mangi Lal Jat, Mirasol Pampolino, Talatam Satyanarayana, Sudarshan Dutta, and Anil Kumar. "Nutrient Management in Wheat: Current Scenario, Improved Strategies and Future Research Needs in India." *Journal of Wheat Research* 4, no. 1 (2013): 1–10.

Makate, Clifton, Marshall Makate, and Nelson Mango. "Wealth-Related Inequalities in Adoption of Drought-Tolerant Maize and Conservation Agriculture in Zimbabwe." *Food Security* 11 (2019): 881–96.

Manassah, Jamal T., and Ernest J. Briskey, eds. *Advances in Food-Producing Systems for Arid and Semiarid Lands Part A.* New York: Academic Press, 1981.

Mann, Charles C. *The Wizard and the Prophet: Two Groundbreaking Scientists and Their Conflicting Visions of the Future of Our Planet.* London: Picador, 2018.

Mann, Charles K. "Packages of Practices; A Step at a Time with Clusters?" Paper presented at the joint meetings of the American Agricultural Economics Association and the Western Agricultural Economics Association, San Diego, CA, July 31–August 3, 1977.

Mastenbroek, A., T. Gumucio, J. Nakanwagi, and C. Kawuma. *Community Based Risk Spectrum Analysis in Uganda Male and Female Livelihood Risks and Barriers to Uptake of Drought Tolerant Maize Varieties.* Working Paper No. 318. Wageningen: CGIAR Research Program on Climate Change, Agriculture and Food Security, 2020.

Matchett, Karin Elizabeth. "Untold Innovation: Scientific Practice and Corn Improvement in Mexico, 1935–1965." PhD diss., University of Minnesota, 2002.

Mazid, Ahmed, Mesut Keser, Koffi N. Amegbeto, Alexey Morgounov, Ahmet Bagci, Kenan Peker, Mustafa Akin, et al. "Measuring the Impact of Agricultural Research: The Case of New Wheat Varieties in Turkey." *Experimental Agriculture* 51, no. 2 (2015): 161–78.

McGuire, Shawn J. "Path-Dependency in Plant Breeding: Challenges Facing Participatory Reforms in the Ethiopian Sorghum Improvement Program." *Agricultural Systems* 96, no. 1–3 (2008): 139–49.

Meyer, Ryan. "The Public Values Failures of Climate Science in the US." *Minerva* 49 (2011): 47–70.

Michaels, Patrick J. "The Response of the 'Green Revolution' to Climatic Variability." *Climatic Change* 4 (1982): 255–71.

Minhas, B. S., and T. N. Srinivasan. "New Agricultural Strategy Analysed." *Yojana* 10 (1966): 20–24.

Mishra, Srijit, A. Ravindra, and Ced Hesse. *Rainfed Agriculture: For an Inclusive, Sustainable and Food Secure India.* IIED Briefing. London: International Institute for Environment and Development, 2013.

Mohammadi, Reza. "Breeding for Increased Drought Tolerance in Wheat: A Review." *Crop and Pasture Science* 69, no. 3 (2018): 223–41.

Mohan, S. T., P. K. Das, and H. K. Jain. "A Note on Adaptation in Wheat." In *Breeding Researches in Asia and Oceania: Proceedings of the Second General Congress of the Society for the Advancement of Breeding Researches in Asia and Oceania, February 22–28, 1973,* edited by S. Ramanujam and R. D. Iyer, 1123–24. New Delhi: Indian Society of Genetics and Plant Breeding, 1974.

Monneveux, Philippe, Ruilian Jing, and Satish C. Misra. "Phenotyping for Drought Adaptation in Wheat Using Physiological Traits." *Frontiers in Physiology* 3, no. 429 (2012): 1–12.

Mruthyunjaya and P. Ranjitha. "The Indian Agricultural Research System: Structure, Current Policy Issues, and Future Orientation." *World Development* 26 (1998): 1089–101.

Munshi, Kaivan. "Social Learning in a Heterogeneous Population: Technology Diffusion in the Indian Green Revolution." *Journal of Development Economics* 73, no. 1 (2004): 185–213.

Myren, Delbert T. "The Rockefeller Foundation Program in Corn and Wheat in Mexico." In *Subsistence Agriculture and Economic Development*, edited by Clifton R. Wharton Jr., 439–53. London: Aldine, 1969.

Nagarajan, S. "Can India Produce Enough Wheat Even by 2020?" *Current Science* 89, no. 9 (2005): 1467–71.

Nagarajan, S., Gyanendra Singh, and B. S. Tyagi, eds. *Wheat Research Needs beyond 2000 AD*. New Delhi: Narosa, 1998.

Naik, K. C. "Inaugural Address." In *Proceedings of the Seminar on Drought, Dec. 26th to 29th, 1968*, 1–5. U.A.S. Res. Series No. 14. Hebbal, Bangalore: University of Agricultural Sciences, 1972.

Narayanan, Sudha. "Food Security in India: The Imperative and Its Challenges." *Asia and the Pacific Policy Studies* 2, no. 1 (2015): 197–209.

National Resources Institute, University of Greenwich. *End of Program Evaluation for AGRA Africa's Seed System Programs: Evaluation Report*. National Resources Institute, July 2019.

Nelson, Rebecca, Richard Coe, and Bettina I. Haussmann. "Farmer Research Networks as a Strategy for Matching Diverse Options and Contexts in Smallholder Agriculture." *Experimental Agriculture* 55, no. S1 (2016): 125–44.

Nene, Y. L. "Plant Protection: A Key to Maintaining Present Gains in Food Production." Public lecture at Cornell Workshop on Some Emerging Issues Accompanying Recent Breakthroughs in Food Production, Ithaca, NY, March 31, 1970.

Newport, Danielle, David B. Lobell, Balwiner-Singh, Amit K. Srivastava, Preeti Rao, Maanya Umashaanker, Ram K. Malik, Andrew McDonald, and Meha Jain. "Factors Constraining Timely Sowing of Wheat as an Adaptation to Climate Change in Eastern India." *Weather, Climate, and Society* 12, no. 3 (2020): 515–28.

Niosi, Jorge, Paolo Saviotti, Bertrand Bellon, and Michael Crow. "National Systems of Innovation: In Search of a Workable Concept." *Technology in Society* 15, no. 2 (1993): 207–27.

Oasa, Edmund K. "The International Rice Research Institute and the Green Revolution: A Case Study on the Politics of Agricultural Research." PhD diss., University of Hawaii, 1981.

Oasa, Edmund K. "The Political Economy of International Agricultural Research: A Review of CGIAR's Response to Criticisms of the 'Green Revolution.'" In *Green Revolution Revisited: Critique and Alternatives*, vol. 2, edited by Bernhard Glaeser, 9–41. Abingdon, UK: Routledge, 2011.

Oficina de Estudios Especiales. *Fourth Annual Progress Report: September 1, 1953–August 31, 1954*. Washington, DC: Rockefeller Foundation, 1954.

Ollenburger, Mary, Todd Crane, Katrien Descheemaeker, and Ken E. Giller. "Are Farmers Searching for an African Green Revolution? Exploring the Solution Space for Agricultural Intensification in Southern Mali." *Experimental Agriculture* 55, no. 2 (2019): 288–310.

Olsson, Tore C. *Agrarian Crossings: Reformers and the Remaking of the US and Mexican Countryside*. Princeton, NJ: Princeton University Press, 2020.

Orr, Alastair. "Why Were So Many Social Scientists Wrong about the Green Revolution? Learning from Bangladesh." *Journal of Development Studies* 48, no. 11 (2012): 1565–86.

Ozgediz, Selcuk. *The CGIAR at 40:Institutional Evolution of the World's Premier Agricultural Research Network*. Washington, DC: CGIAR Fund Office, 2012.

Parayil, Govindan. "The Green Revolution in India: A Case Study of Technological Change." *Technology and Culture* 33, no. 4 (1992): 737–56.

Parikh, Kirit S. "HYV Fertilisers: Synergy or Substitution: Implications for Policy and Prospects for Agricultural Development." *Economic and Political Weekly* 13, no. 12 (March 25, 1978): A2–A8.

Patel, Raj. "The Long Green Revolution." *Journal of Peasant Studies* 40, no. 1 (2013): 1–63.

Pathak, H., B. B. Panda, A. K. Nayak, eds. *Bringing Green Revolution to Eastern India: Experiences and Expectations*. Cuttack: ICAR-National Rice Research Institute, 2019.

Pawley, Emily. *The Nature of the Future: Agriculture, Science, and Capitalism in the Antebellum North*. Chicago: University of Chicago Press, 2020.

Perkins, John H. *Geopolitics and the Green Revolution: Wheat, Genes, and the Cold War*. Oxford: Oxford University Press, 1997.

Pielke, Roger, and Björn-Ola Linnér. "From Green Revolution to Green Evolution: A Critique of the Political Myth of Averted Famine." *Minerva* 57 (2019): 265–91.

Piesse, Mervyn. *Hunger amid Abundance: The Indian Food Security Enigma*. Future Directions International, 2019. https://apo.org.au/node/226011.

Pistorius, Robin. *Scientists, Plants and Politics: A History of the Plant Genetic Resources Movement*. Washington, DC: International Plant Genetic Resources Institute, 1997.

Prasad, C. Shambu. "Constructing Alternative Socio-technical Worlds: Re-imagining RRI through SRI in India." *Science, Technology and Society* 25, no. 2 (2020): 291–307.

Rahman, Abdur, J. D. Hayes, and C. A. Foster. *Manual of Wheat Breeding Proce-*

dures. New York: Food and Agriculture Organization of the United Nations, 1987.

Raina, Rajeswari S. "But Why? Towards Agricultural Science Policy in India." Paper presented at the Atlanta Conference on Science and Innovation Policy, Atlanta, GA, September 15–17, 2011.

Raina, Rajeswari S. "Institutional Strangleholds: Agricultural Science and the State in India." In *Shaping India: Economic Change in Historical Perspective*, edited by D. Narayana and Raman Mahadevan, 99–123. Abingdon, UK: Routledge, 2011.

Raina, Rajeswari S. "Institutions and Organisations Enabling Reforms in Indian Agricultural Research and Policy." *International Journal of Technology Management and Sustainable Development* 2, no. 2 (2003): 97–116.

Raina, Rajeswari S. "Science for a New Agricultural Policy." In *Vicissitudes of Agriculture in the Fast Growing Indian Economy: Challenges, Strategies and the Way Forward*, edited by C. Ramasamy and K. R. Ashok, 252–66. New Delhi: Academic Foundation, 2016.

Raina, Rajeswari S., K. Joseph, E. Haribabu, and Ramesh Kumar. *Agricultural Innovation Systems and the Co-evolution of Exclusion in India*. Working Paper SIID 07/2009. Sheffield, UK: Sheffield Institute for International Development, 2009.

Raina, Rajeswari S., A. Ravindra, M. V. Ramachandrudu, and S. Kiran. *Reviving Knowledge: India's Rainfed Farming, Variability and Diversity*. Briefing Paper 17307IIED, International Institute for Environment and Development, August 2015.

Rajaram, S., N. E. Borlaug, and M. Van Ginkel. "CIMMYT International Wheat Breeding." In *Bread Wheat: Improvement and Production*, edited by Byrd C. Curtis, Sanjaya Rajaram, and Helena Gómez Macpherson, 103–17. Rome: Food and Agriculture Organization of the United Nations, 2002.

Rajaram, S., and S. Ceccarelli. "International Collaboration in Cereal Breeding." In *Wheat: Prospects for Global Improvement: Proceedings of the 5th International Wheat Conference, 10–14 June 1996, Ankara, Turkey*, edited by H.-J. Braun, F. Altay, W. E. Kronstad, S. P. S. Beniwal, and A. McNab, 533–37. New York: Springer, 1997.

Ram, J., O. P. Jain, and B. R. Murty. "Stability of Performance of Some Varieties and Hybrid Derivatives in Rice under High Yielding Varieties Programme." *Indian Journal of Genetics and Plant Breeding* 30, no. 1 (1970): 187–98.

Ramadas, Sendhil, T. M. Kiran Kumar, and Gyanendra Pratap Singh. "Wheat Production in India: Trends and Prospects." In *Recent Advances in Grain Crops Research*, edited by Farooq Shah, Zafar Khan, Amjad Iqbal, Metin Turan and Murat Olgun, 89–104. Rijeka: IntechOpen, 2020. https://www.intechopen.com/chapters/67311.

Ramsden, Edmund. "Carving Up Population Science: Eugenics, Demography and the Controversy over the 'Biological Law' of Population Growth." *Social Studies of Science* 32, no. 5–6 (2002): 857–99.

Randhawa, N. S. *Agricultural Research in India: An Overview of Its Organization, Management and Operations.* New York: Food and Agriculture Organization of the United Nations, 1987.

Rao, N. Ganga Prasada, and G. Harinarayana. "Phenotypic Stability of Hybrids and Varieties in Grain Sorghum." *Current Science* 38, no. 4 (1969): 97–98.

Rao, V. K. R. V. *Food, Nutrition and Poverty in India.* New Delhi: Vikas, 1982.

Rapley, Rob, writer. *American Experience.* "The Man Who Tried to Feed the World." Directed by Rob Rapley. Aired April 21, 2020, in broadcast syndication, WGBH Boston.

Redclift, Michael. "Production Programs for Small Farmers: Plan Puebla as Myth and Reality." *Economic Development and Cultural Change* 31, no. 3 (1983): 551–70.

Reynolds, Matthew P., ed. *Climate Change and Crop Production.* Wallingford, UK: CABI, 2010.

Reynolds, Matthew, Alistair Pask, and Debra Mullan, eds. *Physiological Breeding I: Interdisciplinary Approaches to Improve Crop Adaptation.* Mexico City: CIMMYT, 2012.

Rockefeller Foundation. *Annual Report.* New York: Rockefeller Foundation, 1959.

Rockefeller Foundation. *Indian Agricultural Program: Director's Annual Report. January 15–September 15, 1957.* New York: Rockefeller Foundation, 1957.

Rockefeller Foundation. *Rockefeller Foundation Program in the Agricultural Sciences: Annual Report, 1959–60.* New York: Rockefeller Foundation, 1960.

Rockefeller Foundation. *Rockefeller Foundation Program in the Agricultural Sciences: Annual Report, 1963–4.* New York: Rockefeller Foundation, 1964.

Rockefeller Foundation. *Rockefeller Foundation Program in the Agricultural Sciences: Annual Report, 1964–5.* New York: Rockefeller Foundation, 1965.

Rockefeller Foundation. *Rockefeller Foundation Program in the Agricultural Sciences: Progress Report: Towards the Conquest of Hunger, 1965–6.* New York: Rockefeller Foundation, 1966.

Röling, Niels. "Pathways for Impact: Scientists' Different Perspectives on Agricultural Innovation." *International Journal of Agricultural Sustainability* 7, no. 2 (2009): 83–94.

Romagosa, I., and P. N. Fox. "Genotype × Environment Interaction and Adaptation." In *Plant Breeding: Principles and Prospects,* edited by M. D. Hayward, N. O. Bosemark, and I. Romagosa, 373–90. London: Chapman and Hall, 1993.

Rosielle, A. A., and J. Hamblin. "Theoretical Aspects of Selection for Yield in Stress and Non-stress Environment." *Crop Science* 21, no. 6 (1981): 943–46.

Roy, N. N., and B. R. Murty. "Response to Selection for Wide Adaptation in Bread Wheat." *Current Science* 36 (1967): 481–82.

Roy, N. N., and B. R. Murty. "A Selection Procedure in Wheat for Stress Environment." *Euphytica* 19 (1970): 509–21.

Sagar, Vidya. "Fertiliser Use Efficiency in Indian Agriculture." *Economic and Political Weekly* 30, no. 52 (December 30, 1995): A160–80.

Saha, Madhumita. "State Policy, Agricultural Research and Transformation of Indian Agriculture with Reference to Basic Food-Crops, 1947–75." PhD diss., Iowa State University, 2012.

Salvi, S., O. Porfiri, and S. Ceccarelli. "Nazareno Strampelli, the 'Prophet' of the Green Revolution." *Journal of Agricultural Science* 151, no. 1 (2013): 1–5.

Samberg, Leah, James S. Gerber, Navin Ramankutty, Mario Herrero, and Paul C. West. "Subnational Distribution of Average Farm Size and Smallholder Contributions to Global Food Production." *Environmental Research Letters* 11, no. 12 (2016): article 124010.

Schmalzer, Sigrid. *Red Revolution, Green Revolution: Scientific Farming in Socialist China.* Chicago: University of Chicago Press, 2016.

Schurman, Rachel. "Micro(soft) Managing a 'Green Revolution' for Africa: The New Donor Culture and International Agricultural Development." *World Development* 112 (2018): 180–92.

Schut, Marc, Laurens Klerkx, Murat Sartas, Dieuwke Lamers, Mariette McCampbell, Ifeyinwa Ogbonna, Pawandeep Kaushik, Kwesi Atta-Krah, and Cees Leeuwis. "Innovation Platforms: Experiences with Their Institutional Embedding in Agricultural Research for Development." *Experimental Agriculture* 52, no. 4 (2016): 537–61.

Scott, James C. *Seeing Like a State.* New Haven, CT: Yale University Press, 1998.

Sen, Amartya. "Ingredients of Famine Analysis: Availability and Entitlements." *Quarterly Journal of Economics* 96, no. 3 (1981): 433–64.

Sen, Bandhudas. *The Green Revolution in India: A Perspective.* Hoboken, NJ: John Wiley and Sons, 1974.

Sen, S. R. *Modernising Indian Agriculture: Report on the Intensive Agricultural District Programme, 1960–68*, vol. 1. New Delhi: Ministry of Food, Agriculture, Community Development and Cooperation, 1969.

Shankar, P. S. Vijay. "Towards a Paradigm Shift in India's Rainfed Agriculture." *Innovation and Development* 1, no. 2 (2011): 321–22.

Sheldrake, Rupert. "Setting Innovation Free in Agriculture." In *Rethinking Food and Agriculture: New Ways Forward*, edited by Amir Kassam and Laila Kassam, 1–30. Duxford, UK: Woodhead, 2021.

Shiferaw, Bekele, Melinda Smale, Hans-Joachim Braun, Etienne Duveiller, Mathew Reynolds, and Geoffrey Muricho. "Crops That Feed the World 10. Past Successes and Future Challenges to the Role Played by Wheat in Global Food Security." *Food Security* 5 (2013): 291–317.

Shiva, Vandana. *The Violence of the Green Revolution: Third World Agriculture, Ecology, and Politics*. London: Zed Books, 1991.

Siegel, Benjamin Robert. *Hungry Nation: Food, Famine, and the Making of Modern India*. Cambridge: Cambridge University Press, 2018.

Sikka, S. M., and K. B. L. Jain. "A Study of the Differential Response of Some Varieties of Wheat to Different Environmental and Cultural Conditions. II. Performance under Different Levels of Soil Fertility and Sowing Dates." *Indian Journal of Agronomy* 4 (1960): 154–63.

Simmonds, N. W. "Genotype (*G*), Environment (*E*) and *GE* Components of Crop Yields." *Experimental Agriculture* 17, no. 4 (1981): 355–62.

Simmonds, N. W. "Selection for Local Adaptation in a Plant Breeding Programme." *Theoretical and Applied Genetics* 82 (1991): 363–67.

Sinclair, Fergus, and R. I. C. Coe. "The Options by Context Approach: A Paradigm Shift in Agronomy." *Experimental Agriculture* 55, no. S1 (2019): 1–13.

Singh, Panjab. *Final Report of the Working Group on Agro-climatic Zonal Planning Including Agriculture Development in North-Eastern India for XI Five Year Plan (2007–12)*, vol. 1, *Main Report*. New Delhi: Planning Commission, Government of India, 2006.

Singh, R. P., and R. C. Agrawal. "Improving Efficiency of Seed System by Appropriating Farmer's Rights in India through Adoption and Implementation of Policy of Quality Declared Seed Schemes in Parallel." *MOJ Ecology & Environmental* Sciences 3 (2018): 387–91.

Singh, Sultan, and I. S. Pawar. *Trends in Wheat Breeding*. New Delhi: CBS Publishers and Distributors, 2006.

Sisodia, N. S. "Photoinsensitivity in Wheat Breeding." In *Genetics and Wheat Improvement: Proceedings of the 1st National Seminar on Genetics and Wheat Improvement*, edited by Akshey Kumar Gupta, 172–74. New Delhi: Oxford and IBH, 1977.

Sivaraman, B. "Scientific Agriculture Is Neutral to Scale: The Fallacy and the Remedy." Speech to the 26th Annual Conference of Indian Society of Agricultural Statistics, Kalyani, December 27, 1972.

Skilbeck, D., G. Barbero, C. Bower, E. D. Carter, G. J. Koopman, I. Abu Sharr, and G. van Poorten. *Research Review Mission to the Near East and North Africa: Report to the Technical Advisory Committee of the Consultative Group on International Agricultural Research*. Rome: Food and Agriculture Organization of the United Nations, 1973.

Smith, Elta C. "Governing Rice: The Politics of Experimentation in Global Agriculture." PhD diss., Harvard University, 2008.

Snapp, Sieglinde. "A Mini-review on Overcoming a Calorie-Centric World of Monolithic Annual Crops." *Frontiers in Sustainable Food Systems* 4 (2020): article 540181.

Soto Laveaga, Gabriela. "The Socialist Origins of the Green Revolution: Pandurang Khankhoje and Domestic 'Technical Assistance.'" *History and Technology* 36, no. 3–4 (2020): 337–59.

Spielman, David J., and Rajul Pandya-Lorch, eds. *Proven Successes in Agricultural Development: A Technical Compendium to Millions Fed.* Washington, DC: International Food Policy Research Institute, 2010.

Spitz, Pierre. "Green Revolution Re-examined in India." In *Green Revolution Revisited: Critique and Alternatives*, vol. 2, edited by Bernhard Glaeser, 42–58. Abingdon, UK: Routledge, 2011.

Stakman, E. C., Richard Bradfield, and Paul C. Mangelsdorf. *Campaigns against Hunger.* Cambridge, MA: Belknap Press of Harvard University Press, 1967.

Stone, Glenn Davis. "Commentary: New Histories of the Indian Green Revolution." *Geographical Journal* 185, no. 2 (2019): 243–50.

Stone, Glenn Davis, and Dominic Glover. "Disembedding Grain: Golden Rice, the Green Revolution, and Heirloom Seeds in the Philippines." *Agriculture and Human Values* 34 (2016): 1–16.

Subramanian, Kapil. "Revisiting the Green Revolution: Irrigation and Food Production in Twentieth-Century India." PhD diss., King's College London, 2015.

Sumberg, James, Dennis Keeney, and Benedict Dempsey. "Public Agronomy: Norman Borlaug as 'Brand Hero' for the Green Revolution." *Journal of Development Studies* 48, no. 11 (2012): 1587–1600.

Sumberg, James, and David Reece. "Agricultural Research through a 'New Product Development' Lens." *Experimental Agriculture* 40, no. 3 (2004): 295–314.

Swaminathan, M. S. "The Impact of Dwarfing Genes on Wheat Production." *Journal of the IARI Post-graduate School* 3 (1965): 57–62.

Swaminathan, M. S. *Science and Sustainable Food Security: Selected Papers of M. S. Swaminathan.* Bengaluru: Indian Institute of Science Press, 2009.

Swaminathan, M. S. *Wheat Revolution: A Dialogue.* New Delhi: Macmillan India, 1993.

Takeshima, Hiroyuki. "Geography of Plant Breeding Systems, Agroclimatic Similarity, and Agricultural Productivity: Evidence from Nigeria." *Agricultural Economics* 50, no. 1 (2019): 67–78.

Tandon, J. P., and M. V. Rao. "Organisation of Wheat Research in India and Its Impact." In *Twenty-Five Years of Co-ordinated Wheat Research, 1961–1986*, edited by J. P. Tandon and A. P. Sethi, 1–33. New Delhi: Indian Agricultural Research Institute, 1986.

Thompson, John, and Ian Scoones. "Addressing the Dynamics of Agri-Food Systems: An Emerging Agenda for Social Science Research." *Environmental Science and Policy* 12, no. 4 (2009): 386–97.

Tigerstedt, P. M. A. "Adaptation, Variation and Selection in Marginal Areas." *Euphytica* 77 (1994): 171–74.

Traxler, Greg, and Derek Byerlee. "Linking Technical Change to Research Effort: An Examination of Aggregation and Spillovers Effects." *Agricultural Economics* 24, no. 3 (2001): 235–46.

Trouiller, Patrice, Piero Olliaro, Els Torreele, James Orbinski, Richard Laing, and Nathan Ford. "Drug Development for Neglected Diseases: A Deficient Market and a Public-Health Policy Failure." *Lancet* 359, no. 9324 (2002): 2188–94.

UNICEF. *Malnutrition Rates Remain Alarming: Stunting Is Declining Too Slowly While Wasting Still Impacts the Lives of Far Too Many Young Children.* May 2018. https://www.unicef.org/rosa/reports/levels-and-trends-child -malnutrition.

US Agency for International Development. *Synthesis Report: Review of Successful Scaling of Agricultural Technologies.* Washington, DC: USAID Bureau for Food Security, 2017.

US Department of Agriculture Foreign Agricultural Service. *Grain and Feed Annual: Colombia.* April 1, 2020. Report No. CO2020-0001. https://www .fas.usda.gov/data/colombia-grain-and-feed-annual-5.

US National Committee for the International Biological Program. *Preliminary Framework of the U.S. Program of the IBP.* Washington, DC: National Academies Press, 1965.

US Patent Office. *Report of the Commissioner of Patents for the Year 1853: The Agricultural Portion of the Report of that Office for the Year 1853.* Washington, DC: United States Patent Office, 1854.

Vaidyanathan, A. "HYV and Fertilisers: Synergy or Substitution?: A Comment." *Economic and Political Weekly* 13, no. 25 (June 24, 1978): 1031–35.

Verma, M. M., G. S. Chahal, and B. R. Murty. "Limitations of Conventional Regression Analysis a Proposed Modification." *Theoretical and Applied Genetics* 53 (1978): 89–91.

Villareal, R. L. and A. R. Klatt, eds. *Wheats for More Tropical Environments: A Proceedings of the International Symposium.* Mexico City: CIMMYT, 1985.

Visser, Jozef. "Down to Earth: A Historical-Sociological Analysis of the Rise and Fall of 'Industrial' Agriculture and of the Prospects for the Re-rooting of Agriculture from the Factory to the Local Farmer and Ecology." PhD diss., Wageningen University and Research, 2010.

Wade, Nicholas. "Green Revolution (I): A Just Technology, Often Unjust in Use." *Science* 186, no. 4169 (1974): 1093.

Waines, J. Giles, and Bahman Ehdaie. "Domestication and Crop Physiology: Roots of Green-Revolution Wheat." *Annals of Botany* 100, no. 5 (2007): 991–98.

Weisskopf, Thomas E. "Dependence and Imperialism in India." *Review of Radical Political Economics* 5, no. 1 (1973): 53–96.

Wellhausen, E. J. "The Urgency of Accelerating Production on Small Farms." In *Strategies for Increasing Agricultural Production on Small Holdings*, edited

by Delbert T. Myren, 5–10. Mexico City: CIMMYT, 1970. https://ufdc.ufl
.edu/UF00075706/00001/3x.

Westad, Odd Arne. *The Global Cold War: Third World Interventions and the Making of Our Times.* Cambridge: Cambridge University Press, 2005.

Williams, R. *The Khanna Study, the Rockefeller Foundation and Population Control in India.* Paper presented at the Warwick-Colombia-LSE Global History Workshop, Coventry, UK, March 9, 2012.

Witcombe, John, Daljit Virk, and John Farrington, eds. *Seeds of Choice: Making the Most of New Varieties for Small Farmers.* London: Intermediate Technology Publications, 1998.

World Bank. *Implementation Completion Report, India, National Agricultural Research Project II.* Report no. 16612. Washington, DC: World Bank, 1997. http://documents.worldbank.org/curated/en/1997/05/731937/india-second -national-agricultural-research-project.

World Bank. *Report and Recommendation of the President of the International Bank for Reconstruction and Development to the Executive Directors on a Proposed Credit to India for the Second National Agricultural Research Project.* Washington, DC: World Bank, 1985.

Xynias, Ioannis N., Ioannis Mylonas, Evangelos G. Korpetis, Elissavet Ninou, Aphrodite Tsaballa, Ilias D. Avdikos, and Athanasios G. Mavromatis. "Durum Wheat Breeding in the Mediterranean Region: Current Status and Future Prospects." *Agronomy* 10, no. 3 (2020): 432.

Yadav, Rajbir, S. S. Singh, Neelu Jain, G. P. Singh, and K. V. Prabhu. "Wheat Production in India: Technologies to Face Future Challenges." *Journal of Agricultural Science* 2, no. 2 (2010): 164–73.

Yapa, Lakshman. "What Are Improved Seeds? An Epistemology of the Green Revolution." *Economic Geography* 69, no. 3 (1993): 254–73.

Yasin, B. Abrar, M. Ram, Shubhra Singh, and B. A. Wani. "Genetic Improvement in Yield, Yield Attributes and Leaf Rust Resistance in Semi-dwarf Wheat Varieties Developed in India from Last 40 Years." *International Journal of Agricultural Research* 6, no. 10 (2011): 747–53.

Yates, F., and W. G. Cochran. "The Analysis of Groups of Experiments." *Journal of Agricultural Science* 28, no. 4 (1938): 556–80.

Yediay, F. E., E. E. Andeden, F. S. Baloch, A. Börner, B. Kilian, and H. Özkan. "The Allelic State at the Major Semi-dwarfing Genes in a Panel of Turkish Bread Wheat Cultivars and Landraces." *Plant Genetic Resources* 9, no. 3 (2011): 423–29.

INDEX

Note: References in *italic* refer to figures and tables.